An Innovative Approach To Landfill Engineering

Revolutionizing The Landfill Industry

By

Reg Renaud

Th e purpose of this book is to educate. It is sold with the understanding that the
author and publisher shall have neither liability nor responsibility for any injury
caused or alleged to be caused directly or indirectly by the information in this book.
While every eff ort has been made to ensure the accuracy, its contents should not
be construed as site specifi c advice. Each landfi ll has its own characterizations
that may require adjustments to comments made in general in this book.

Order this book online at www.trafford.com
or email orders@trafford.com

Most Trafford titles are also available at major online book retailers.

Print information available on the last page.

ISBN: 978-1-4251-6862-9 (sc)

Because of the dynamic nature of the Internet, any web addresses or links contained in
this book may have changed since publication and may no longer be valid. The views
expressed in this work are solely those of the author and do not necessarily reflect the
views of the publisher, and the publisher hereby disclaims any responsibility for them.

Any people depicted in stock imagery provided by Getty Images are models, and such images
are being used for illustrative purposes only.
Certain stock imagery © Getty Images.

Trafford rev. 08/27/2019

 www.trafford.com
North America & international
toll-free: 1 888 232 4444 (USA & Canada)
fax: 812 355 4082

Table of Contents

Table of Contents
(continued)

Table of Contents
(continued)

Table of Contents
(continued)

Figures

Table of Contents

(continued)

Table of Contents
(continued)

Table of Contents
(continued)

Acknowledgements

In writing this book I think I tapped the various talents of all of my family and friends in one way or another and I am grateful to them. I would like to give special thanks to some of the people who put in that extra effort to make this book possible. My long time engineer friend Michael L. Leonard Sr. who has read and edited everything I have written in the last 20 years and he was not shy with a red ink pen. My cousin Nadine Lalich who had to unscramble some of the messes I made with various computer programs, she may never recover. Also my brother–in-law Wes Hoover and his in-depth knowledge of computers and software. Gone are the days when an author hides away in a room and bangs away on a typewriter. It now requires a team of supportive friends.

Preface

Many technical publications relating to municipal solid waste landfills obtained most of their information from work performed by others in the industry. These books have also been written by academics with the assumption that these books will only be read by other academics with basically just findings and conclusions and little information on how the work was actually done.

This book will list few references of other publications due to the fact that most of the information printed on the following pages is from laboratory tests and actual projects performed in the field by the author. Not only will there be many findings and conclusions but also detailed procedures on how these findings were obtained. The author has attempted to provide this information in such a way as to be more useful and used in the solid waste industry.

It is not the intention of the author to be critical of any one person or of consulting firms, but of the industry in general. Many procedures described in this book will be described as they have been performed for decades by consulting firms and landfill operators. They were thought to be the right thing to do at the time, but have been demonstrated to be faulty with today's technology. Getting the industry to change their ways has been like converting someone to change their religious beliefs.

It is hoped that the reader will find this information useful and will put it to good use in keeping our solid waste landfills in compliance and to protect our environment.

An Innovative Approach
To Landfill Engineering

1.0 INTRODUCTION

The design and operation of Municipal Solid Waste (MSW) landfills has changed over the last few decades primarily due to governmental regulations created by environmental impacts from MSW landfills. It appears that once these regulations were in place the methodology used to keep a landfill in compliance with the regulations, began to stagnate and new advances have slowed to a stand still. Change in technology can occur quickly if a need arises, but change in regulations takes a very long time unless death or disaster is eminent or has occurred.

To maintain compliance of current regulations usually requires monitoring of the surface emissions from the landfill, soil gas probes in the surrounding soils and groundwater monitoring wells. When Landfill Gas (LFG) is detected above acceptable levels, a gas collection system is usually designed and installed to extract the LFG in a controlled manner. Most of these collection system designs are based on the readings obtained from these monitoring devices.

The current regulations require most new landfills to have a bottom liner system to prevent migration of leachate from the landfill. These liners were not intended to contain LFG migration, this was to be controlled by the LFG collection system.

Most liner and collection designs on current landfills are based on information from the monitoring readings from the above-mentioned probes, surface readings and specific site conditions and materials used.

The regulations also require that the refuse remain as dry as possible (dry tomb method) to prevent methane production and leachate migration. However, this also prevents the organic materials in the landfill to decompose.

The major flaw in the current method of landfill engineering is that all monitoring and reaction to the compliance requirements are performed at "top of well readings" that provide minimal amount of information as to where the LFG is coming from. A LFG collector provides a cross-section of the gas being produced along the full depth of the collector.

This book will discuss and explore a new approach to landfill design, monitoring, liners, collectors and enhanced bioreaction with the use of in-situ instruments. The instruments and procedures discussed in the following pages will provide the reader with a better understanding of the dynamic and ever-changing conditions inside a MSW landfill.

The first chapter will begin with the history of the development of the cone penetrometer test (CPT) and its in-situ functions. It will also discuss the addition of the Piezo-Penetrometer Test (PPT) with the installation of a pressure transducer.

The subsequent chapters will discuss the discoveries made from these instruments and their contribution to new approaches to landfill engineering and compliance.

Chapter 1

1.1 In Situ Instruments

I cannot count how many times I have heard engineers at landfills say, "I wish I could quickly and cheaply put a pipe in a spot in a landfill take a sample and verify that there is significant amount of LFG present before spending the money on a collector." That is what the Piezo-Penetrometer Test (PPT) can do and provide even more valuable information than a simple pipe. The following is a short explanation of the capabilities of this instrument.

The electronic cone penetrometer is an instrument made of stainless steel with two strain gauge bridges mounted internally. They range from 1.5 inches (10 cm^2) to 1.75 inches in diameter (15 cm^2) and approximately 10 to 12 inches long.

There have been small mini cones (½ to ¾" diameter) developed but they are rare and are not used in landfill investigations. The CPT measures tip resistance and sleeve resistance (useful data for soil type, strength and pile design) as it is hydraulically pushed into the landfill by a 20-ton truck, not drilled.

The electronic cone penetrometer test (CPT) was developed in Holland in the mid 1970's and was very useful in foundation studies in the soft hydro-fills found in that country. In the late 1970's it was introduced to the U.S. but had to be modified for the stiffer soils found in North America. It is often used for pile foundation design. In 1977 a CPT unit, mounted on a World War II US army truck, was shipped from Holland to Houston, Texas where I and a co-worker were assigned the task of making it work.

The truck was not legal to drive on US roads so we had to remove all of the hydraulic system and electronics off the old truck and install it on a new truck. The trick was converting the metric piping into standard fittings, in the 1970's this was a challenge.

With the addition of a pressure transducer in the cone the Piezo-Penetrometer Test was developed in the early 1980's.

At the time, the primary purpose of this device was to measure the pore-pressure in the soil to determine the level of the groundwater. The PPT also measures tip resistance and sleeve resistance the same as a standard CPT. The instrument has inclinometers to determine that the cone is staying vertical during it's advancement through the subsurface or landfill. An encoder is attached to the push rods, which provides very accurate depth control. Some cone manufacturers install thermistors in the cone to correct for temperature influences on the strain gauge bridges as heat increases as the cone is advanced through the soil.

There are basically two configurations of the PPT cone, one is a tip sensing unit which has the pressure transducer placed in the tip and is surrounded by a porous element to allow liquid or gas to enter the transducer chamber.

The second configuration is the side- sensing cone, which has the pressure transducer behind the tip of the cone and the porous element is a ring located behind the tip as well.

Different pressure transducers with many different pressure ranges can be used in either configuration. If low pressures are anticipated, a very sensitive transducer can be used to provide greater resolution. If high pressures are anticipated, a transducer that operates as high as 1,000 psi can be installed.

The tip-sensing cone is very responsive to changing pressures because it operates in the positive mode caused by dynamic forces as the cone is advanced through the soil. However, the porous element on the tip-sensing cone is more vulnerable to damage when being pushed through dense material such as trash.

This configuration is usually used in native soils surrounding the landfill such as pre-qualifying locations for installing perimeter probes. It is also used to detect migrating LFG pressures in native soils and to indicate vacuum pressures from nearby collectors that may have influence outside the landfill. The side-sensing cone is typically used for landfill profiling because it can be pushed through the dense layers of trash without damage.

The side-sensing cone is not as responsive as the tip-sensing cone because it operates from the negative side of the range due to the dynamic vacuum created when the cone is advanced through the trash. To obtain a stabilized reading it is necessary to stop the advancement of the cone and allow the pressures or vacuums to reach a steady state, which is sometimes recorded over time to produce a curve.

Once a profile of an area of a landfill has been developed, it may be desirable to investigate the gas plumes further.

Another instrument that could be used is a thermistor that can be installed on the tip in place of the pressure transducer (some cone companies have both at the same time). The temperature of the gas plumes can be taken to determine the quality of the gas or to indicate a subterranean landfill fire.

If the temperature is above 150 degrees F, it may indicate an underground fire or rapid oxidation, which would produce more CO_2 than methane. If the temperatures are acceptable a soil gas slide-sampling cone can be used to take a sample from a plume and have it analyzed. This is a much slower process but the sample will be very discreet. It is best used to map the migration of gas plumes outside the trash prism, in the vadose zone and in the surrounding soils. This process is helpful if two landfills are adjacent to one another and there is an impacted perimeter probe sitting between them. A gas sample recovered from each landfill and compared to a sample from the perimeter probe may determine which landfill is impacting the probe.

The following diagrams show the side sensing PPT, tip sensing PPT and the basic soil gas slide-sampler.

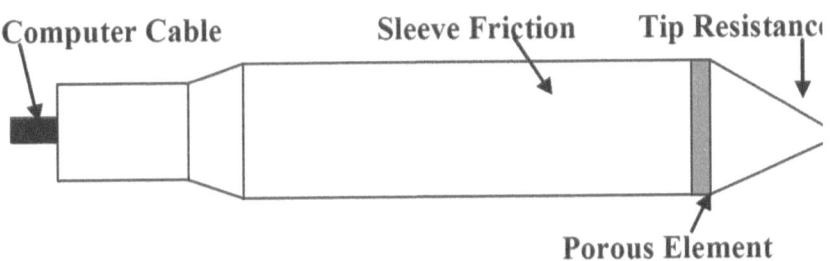

Side- Sensing Piezo-Penetrometer

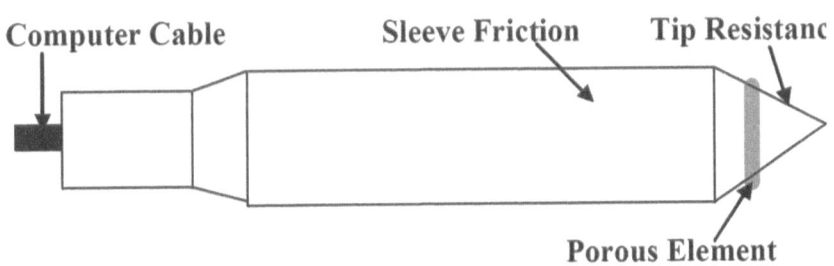

Tip-Sensing Piezo-Penetrometer

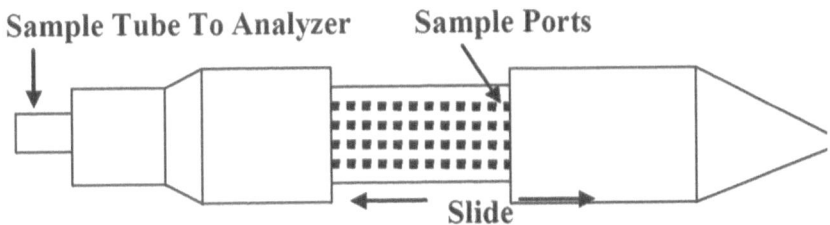

Soil Gas Slide-Sampler

Some cone companies have soil gas sampling capabilities built into the PPT cone as one unit.

Figure 1-1

1.2 The Procedure

It is well known that decomposition of Municipal Solid Waste (MSW) in landfills produces methane gas and other trace toxic gases (e.g. vinyl chloride, etc.). It is necessary to prevent this gas from escaping into the atmosphere to stay in compliance with current Federal and State air pollution prevention regulations. It is also important for landfills located near structures (residential, commercial) to prevent methane from migrating to the structures and causing a potential explosive condition. Finally, uncontrolled landfill gas (LFG) migration at sites with shallow groundwater can result in contamination of the water source, which can cost the landfill owner/operator many millions of dollars for clean-up and abatement.

Historically, as the environmental industry was being developed in the 1980's it was the groundwater that was in eminent danger so groundwater monitoring and extraction wells were developed. As government regulations expanded to include landfills restricting releases of leachate and LFG the landfill operators turned to engineering consulting firms for guidance and advice. However, the only experience the consultants had was with groundwater monitoring wells so they started installing the same type of wells in landfills.

Today, the standard practice for LFG control is to install gas extraction wells or collectors, based on information obtained from grid walk monitoring of gas seepage from the landfill surface and from probes installed at widely spaced intervals around the landfill perimeter. Once it is determined that a gas extraction well or collector is to be placed, it is installed using several drilling methods such as hollow stem auger, bucket/open flight auger or air rotary, or by the horizontal trench method. For vertical wells, the well casing may vary in diameter from 4" to 24" and the depth will also vary.

Most well casings are at least 4" in diameter, in order to accommodate a submersible pump to remove leachate, which enters the collector through the screen and floods the well and stops the flow of gas into the well.

Once this leachate is removed it must then be either treated or disposed of off-site. Vertical collectors are typically screened or slotted from near the bottom to within about 10 feet from the surface because it is generally not known at what level the gas will be located. Installing the screen so close to the surface can result in air infiltration into the gas stream unless the flow rate is restricted by a control valve.

Also, the screen section and surrounding sand or gravel pack can act as a conduit for perched leachate to flow down to the bottom of the landfill where it may impact the groundwater. The main reason these collectors are at least 4" in diameter is to accommodate pumps, because it is not known prior to construction if the collector is being installed through a perched liquid layer or not.

The standard drilling methods cannot typically locate thin layers of perched water, which may impact the collectors after installation. It has been observed during drilling that the refuse appeared to be dry but following completion of the collector liquids were detected in the collector with flooded screens.

To eliminate these problems a method has been developed to determine if the location selected for a collector is suitable for gas extraction and has little leachate, which will affect the production of the collector.

To determine where to install collectors it is necessary to profile the trash prism and outline where the gas pockets are and where perched leachate exists, which should be avoided. It is also necessary to map out the existing vacuum influence from nearby collectors to ensure full coverage of the landfill.

The following outlines the development of the use of the PPT and cone penetrometer test (CPT) for profiling of gas pressures, perched liquid, vacuum zones, and dense cover soil layers in landfills and the use of the acquired data for location and design of LFG collectors. Although the quality of LFG based on the percentage of methane contained in the LFG is useful in flare or co-gen. design and operations it is pressure that is important in landfill compliance.

When gas pressure is measured it is assumed it contains methane however this also means that there has been or there is currently LFG being generated in this location.

It is high pressure that will cause migration to a low-pressure area. Pressure is caused by confinement (free flow does not create pressure) or a higher production rate than the permeability of the waste in the area. Design and construction of LFG collection systems based on this in situ data can be significantly more effective and efficient.

1.3 Piezo-Penetrometer Test Operations

The Piezo-Penetrometer Test (PPT) cone is pushed into the landfill by hydraulic rams mounted on a 20-30 ton truck not drilled in. The truck usually has hydraulic leveling jacks to ensure that the cone starts at the vertical prior to being pushed into the landfill. This also ensures that the full 20-30 ton reaction force is centered on the cone rods. The leveling jacks also provide support for the truck suspension when pulling out the cone rods once the sounding is complete. It can sometimes require more than 20-30 tons of force to pull out the rods, which is more than the truck suspension can take.

Figure 1-2

The cone is advanced by adding 1-meter long push rods pre-threaded with a cable connecting the cone to a computer, which digitally records and displays in real time several parameters continuously during the sounding.

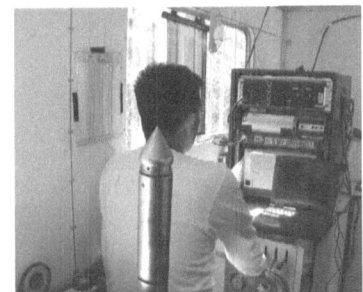

Figure 1-3

The instrument has inclinometers to determine that the cone is staying vertical during advancement through the subsurface or landfill, which adds to the accuracy of the depth control.

11

The depth control of the sounding is very exact (via a wire-line from an encoder attached to each advanced push rod segment).

Depending on the layout of the grid and the depth of each sounding the cone rig can perform between 500 to 1000 feet of sounding per day. The cone rig can typically drive anywhere a pick-up truck can go except for some height limitations (12 feet). The rig also weighs 20-30 tons so some small bridges may need to be supported. Other types of cone vehicles are available for unique conditions such as: tracked vehicles, trailer mounted and backyard portables.

The process is a very clean and safe operation and can operate in any weather condition. In most cases only a four-gas analyzer is used inside the working cab to monitor the air space for methane an H2S. Rod wipers outside the working cab clean the cone rods before they enter the cab. The only sound that is produced is from the truck's engine so it is not obtrusive in sensitive neighborhoods.

Figure 1-4

Following the sounding, the PPT log is printed in the field so it can be used to decide if a LFG collector should be installed in this location. This decision can be made at this time or it may be necessary to wait until the entire grid is completed so that best location can be selected.

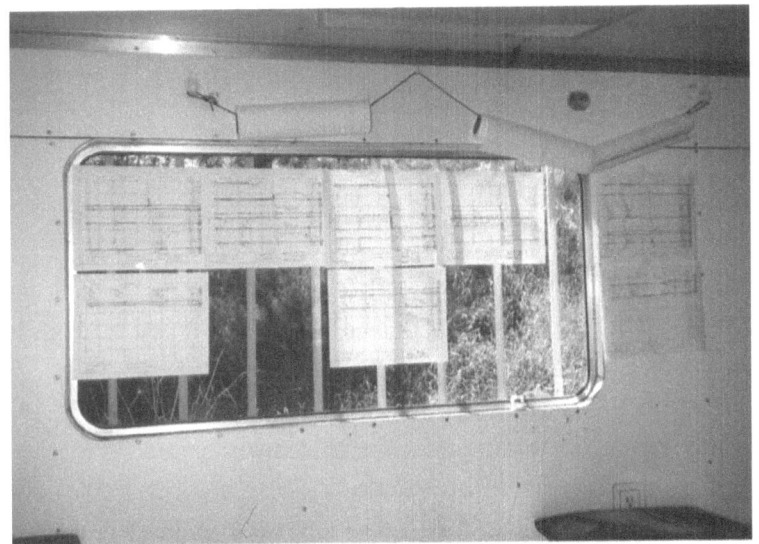

Figure 1-5

No cuttings are produced since the device is pushed into the landfill and not drilled in and the 1 ¾ inch diameter sounding hole that is created can be easily surface sealed with Bentonite chips and hydrated following the sounding.

1.4 PPT Log Interpretations

Figure 1-6 presents a typical PPT sounding log and describes how to read the log. Different cone companies display their data a little differently but the units will be the same as stated in ASTM Standards. Some logs are displayed in black and white and some are in color.

13

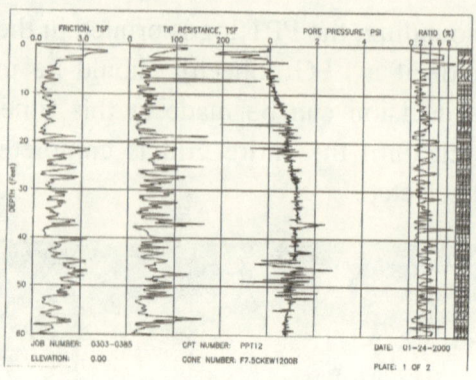

Figure 1-6

The above PPT log is a typical sounding where landfill gas is indicated. The following is a brief description of each column displayed on the log:

Column 1 – Depth of sounding in ten foot intervals. Changing this scale increases or decreases the resolution of the data.

Column 2 – Friction sleeve values as the PPT cone is advanced through the landfill. The values are presented in tons per square foot (tsf). The friction values are useful in determining a friction ratio, which is used to identify the type of material the cone is passing through. It is also an indicator of moist conditions in the landfill. The lower the friction value the higher the moisture content.

Column 3 – Tip Resistance or End Bearing (tsf) values indicate the relative density of the material the cone is penetrating. This value is also used in the friction ratio calculation. The high tip resistance readings can indicate dense layers or daily cover layers and the low tip values usually indicate refuse.

Column 4 – Pore Pressure values (psi) can measure gas pressures, vacuum and liquid head pressure. On the above PPT log, the Pore Pressure values begin to increase in pressure at a depth of 5 feet below ground surface (bgs) and continue to increase to about 1 psi of gas pressure all the way to 60 feet bgs.

14

Column 5 – Friction ratio (%) is calculated by dividing the friction sleeve value by the tip resistance and is presented in a percent. In soils, friction ratios of less than 2% typically indicate sandy or gravelly soil behavior types, and the higher the friction ratio, the more "clay-like" the material (Robertson and Campanella 1988). Moist municipal solid waste has been found to generally have friction ratios greater than 2%.

Column 6 – Lithology is presented by some cone companies with symbols and some companies actually print the soil classification out in words.

The current software programs do not identify or print "refuse", it is usually printed as a clay or a silt based on the friction ratio.

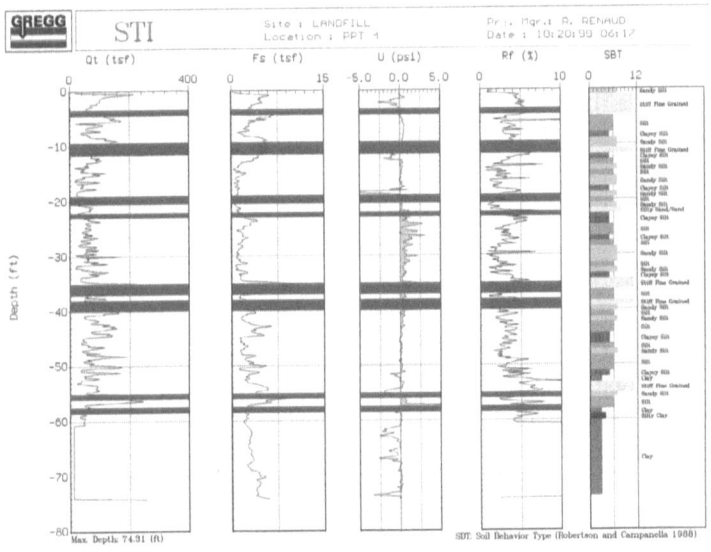

Figure 1-7

This is a color format of a PPT log. Some people think this is a better-looking presentation of the data but the small resolution of each column make interpretations more difficult.

The brown lines across the log were placed by STI, not the Cone company, we are identifying the dense layers.

15

Also STI placed the yellow on the pore pressure column to highlight the gas pressure. Please note the native clay soil at the bottom of the landfill. Here it is easy to determine the bottom of trash.

1.5 Stabilization

To obtain a stabilized reading (i.e. eliminate the affects of cone advancement) it is sometimes necessary to stop the advancement of the cone and allow the pressures or vacuum to reach a steady state, which is recorded over time to produce a curve.

These stabilization tests are typically performed at specified intervals (e.g. 3 to 5 feet) or when significant changes occur in pore-pressure or tip/sleeve resistance readings. An example test result is shown in Figure 1-8.

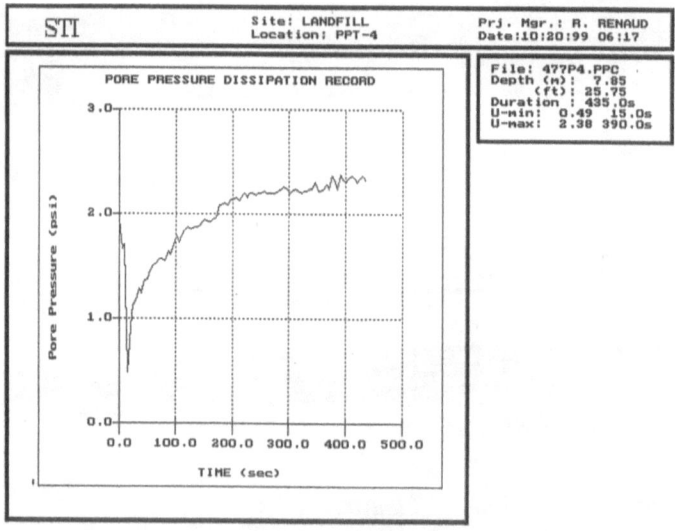

Figure 1-8

The above stabilization curve is a typical log of gas pressure. The initial pressure reading at the start of this test was 1.9 psi.

The cone was retracted 1 foot (note the negative reading as the cone is retracted) to clear the porous element from possible blinding from plastic or other refuse.

Also, if the pressure had dropped and stabilized by about 0.5 psi (equivalent to 12 inches of water column) it would have indicated a liquid head pressure. However, since the pressure reading increased over time, to about 2.4 psi this indicates LFG rather than a liquid.

It is during this stabilization test that it is determined if the pressure being measured is from liquid or gas. Liquid will produce a hydrostatic pressure distribution (see Figure 1-9) based on the depth of the liquid while LFG will produce various pressures, which are independent of the depth.

If the pressure measurements are significantly higher or lower than the hydrostatic pressure could possibly be for the depth, then it is gas pressure that is being measured.

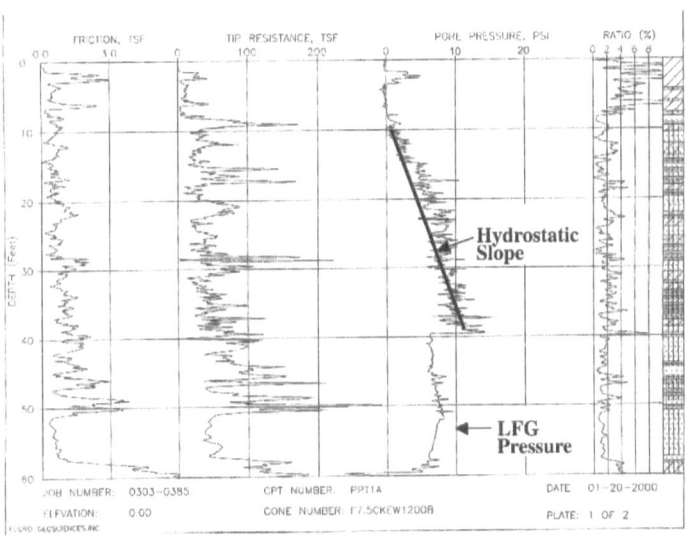

Figure 1-9

17

The PPT log Figure 1-9, indicates a hydrostatic slope from 8 feet bgs to 40 feet bgs. It is historically known that when this area of the landfill was constructed there was an intermediate cap over phase 1, placed at 40 feet bgs.

The second phase of the landfill was filled during the rainy season trapping liquids. Note the gas pressure reading between 40 feet and 60 feet that is trapped under the intermediate cap and the liquid. The horizontal collector in the immediate area and below the 40-foot dense layer was not functioning properly and not recovering the gas causing a possible pathway to a nearby impacted probe. If a vertical collector had been installed at this location it would have flooded and transmitted this liquid to the bottom of the unlined landfill.

Typically, as a PPT sounding in a landfill progresses, the side-sensing transducer indicates a slight apparent/dynamic vacuum (e.g. 1 to 2 psi) as the porous element passes through dry layers of refuse or daily cover soil.

When the sounding is stopped to add another push rod (standard length 1 meter) the vacuum reading typically returns to zero. It is during this pause that a vacuum zone induced by the LFG collection system will be registered as the induced vacuum increases away from the zero line (usually the PPT reading is practically the same as the vacuum gauge reading attached to a nearby collector).

If a PPT is being performed within 100 feet of an existing collector, the vacuum should be increased so that the vacuum will be very apparent if it is encountered in the PPT sounding and assist in determining the zone of vacuum influence.

Figure 1-10 presents a record of measured vacuum confirmed by vacuum gauge readings on a nearby horizontal LFG collector. Note the increased vacuum as the PPT moves closer to the horizontal collector with depth.

18

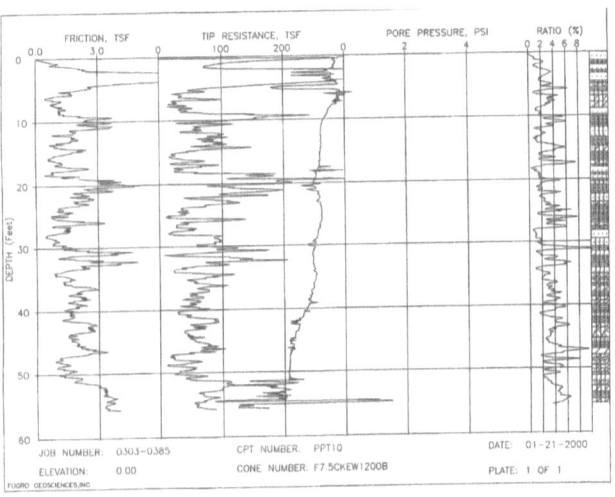

Figure 1-10

If a zone of gas pressure is encountered during the sounding, which is high enough to overcome the small dynamic vacuum, the pore pressure readings will increase and a stabilization test should be performed to measure the static pressure. To clear the porous element and to verify the pressure reading the cone should be pulled up slightly to initiate the stabilization test.

Interpretations of the tip resistance and friction ratio data are made to determine locations of dense daily cover/soil layers, buried access roads or berms, and the bottom of unlined landfills. Sharp increases in tip resistance combined with friction ratios in the range of sandy soils (e.g.< 2%) indicate soil layers within the MSW profile. Typically, gas pockets are found just below a dense soil layer and above a liquid zone. Liquids are frequently found perched above very dense layers.

It has been observed that the average tip resistance in a landfill can give an indication of the density of the refuse and the permeability of the material.

If the PPT is indicating a clay layer when the cone is stopped for a stabilization test, it may require a lot of time for the pressure to stabilize. It is usually more effective if the cone can be stopped in a more sandy layer and more permeable. This information is useful when designing a LFG collection system.

1.6 Data Display and Interpretation

Following a group of soundings performed in an area of a landfill, each of the PPT logs are interpreted where each parameter (i.e. dense layers, liquid and gas layers and vacuum layers) are identified and color coded. The associated data (i.e. location, depth, and thickness of layers) is inputted into Auto Cad and Micro Station files to produce a 3 dimensional profile of the conditions in the area of concern. Also, plan-view and cross sectional displays of the data can be easily generated.

The PPT soundings must be performed in a box type formation to provide depth to a 3-D profile. It is necessary to survey the elevations and location of the surface of each PPT sounding to provide a reference to each of the PPT locations in the 3-D Profile.

The 3-D Profiles or cross-sections and plan views are then evaluated to provide information on:

• Depth of refuse
• Thickness of surface cover soil
• Location and extent of perched liquids
• Location and extent of LFG pressure zones
• Location and extent of dense/daily cover layers
• Understanding the cause or source of LFG migration
• Extent of vacuum influence on and off site
• Where to look for migrating LFG offsite
• Where to place Internal Conduits
• Where to extract discreet LFG samples
• Design of efficient, cost effective collection systems

Following the profiling of a landfill, the data is used to develop a remediation plan to address the gas plumes and other issues required for the landfill to stay in compliance or to recover gas for beneficial uses (e.g. generation of electricity, liquefied gas alternative fuel or the production of hydrogen). With the location of the gas plumes mapped out and with the knowledge of where the perched liquids are located, it is easier to avoid placing collectors through a perched zone which will flood the collector.

One of the most important benefits in using the profiling technique is to prevent installing a collector through a perched liquid zone and providing a conduit for the liquid to migrate to the bottom of the landfill and leak to the groundwater or flood the well. The extraction well screen is placed only where the indicated gas plume is known to be and not in liquid containing zones.

This method eliminates the high cost of pumping hazardous leachate. Typically, the best location to place a collector is just above a liquid layer because that is where LFG will be generated.

1.7 PPT 3-D Profile

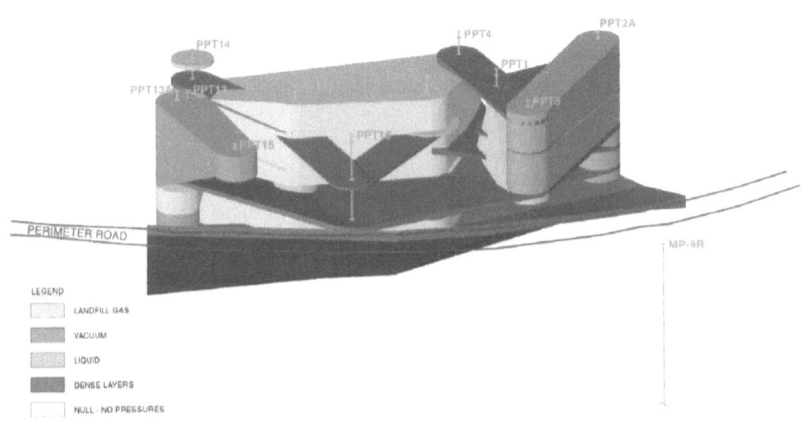

Figure 1-11

21

The above figure is a 3-D Profile developed from a PPT investigation performed at a landfill in Southern California. The profile can be rotated 360 degrees in all directions to provide the best view of the situation. Colors can be turned off to highlight a particular parameter such as LFG (yellow) only.

This profile identified a 30-foot thick layer of liquid (blue) which would have flooded a conventional drilled collector. The profile also identified a gap in the vacuum (pink) coverage in the foreground, which was allowing LFG to migrate towards the perimeter probe. A large volume of LFG (yellow) was indicated beyond the vacuum influence of the nearby collectors.

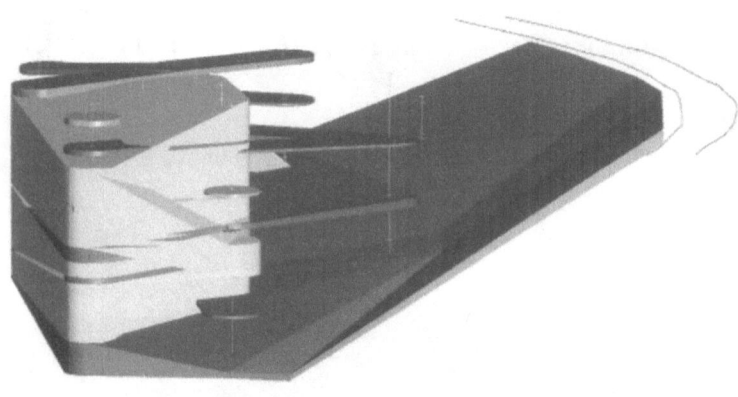

Figure 1-12

This is another view of the previous figure showing a different perspective of the in situ conditions of the landfill. The vacuum (pink) color is turned off. The PPT can identify dense soil layers (brown) such as daily cover or traffic decks. It is typical to find liquid layers perched on top of these dense layers and LFG is typically found trapped under these dense layers.

These dense layers are a double edge sword in that they prevent liquids from migrating to the bottom of the landfill but they also trap LFG under them and cause lateral migration to the edge of the landfill. If it is a canyon fill the LFG will migrate offsite through the native soil. If it is a heaped landfill the LFG can migrate to the slopes and cause surface emissions. As stated above the dense layers will prevent liquids from migrating to the bottom of the landfill until blind conventional drilling to install LFG collectors penetrate these layers and allow the liquid to drain to the bottom of the landfill.

This redistribution of liquids will cause renewed generation of LFG, which may be extracted by the collector unless the collector is flooded by the liquid. The LFG that is not extracted by the collector will be free to migrate.

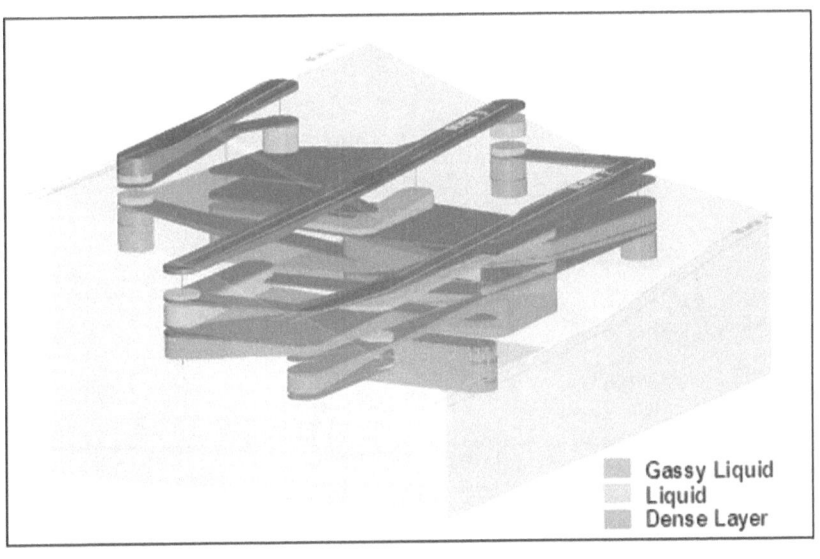

Figure 1-13

The above 3-D profile is an investigation of a section of a landfill that had 41 flooded LFG collectors on three bench roads.

The 3-D process is very helpful when trying to connect the data from various elevations such as on bench roads. The LFG collectors were flooded by over irrigation of the landfill slopes. As the liquid migrates down and saturates the old waste LFG begins to be generated again and is bound up into the liquid creating a condition called gassy liquid, which is indicated above in green. Gassy liquid can create very dangerous conditions in landfills. Very high pore pressure can be created and can affect slope stability. A very fat clay was used as daily cover at this landfill and gas pressures as high as 237 psi was indicated during this investigation.

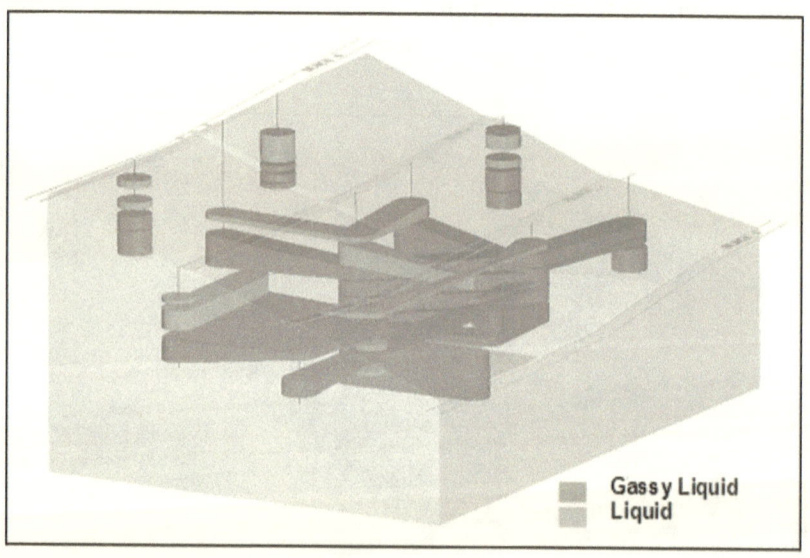

Gassy Liquid
Liquid

Figure 1-14

The figure above is the same profile but with brown color turned off allowing a better view of the liquid and gassy liquid layers. Due to the saturated conditions of the waste prism and the flooded extraction wells no vacuum was indicated.

24

The hydrostatic pressures of the liquid in the waste masked gas pressures. These findings were verified when a backhoe on site dug a 6 foot deep hole on a nearby bench road and artesian like water began to bubble up to the surface and flowed down the bench road. The excavation was quickly backfilled.

1.8 Excess Pore Pressure

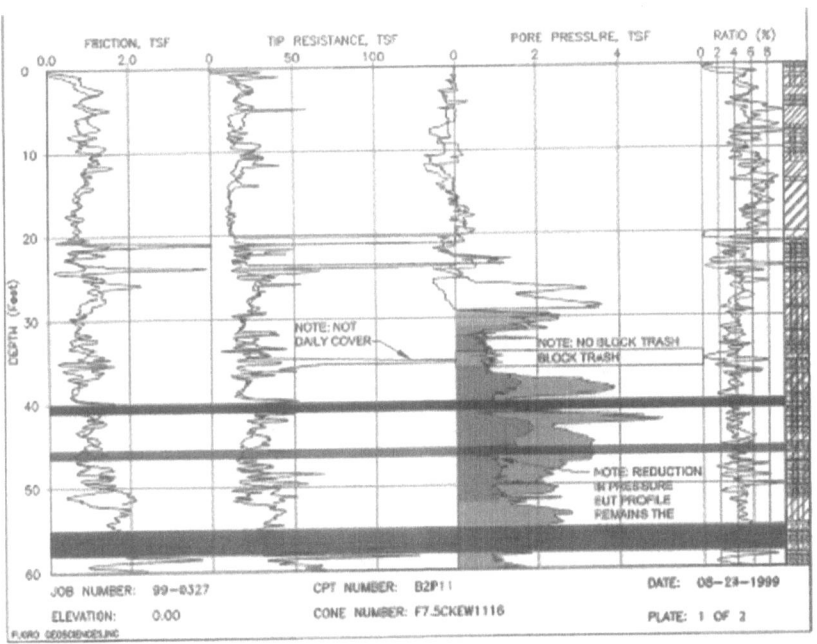

Figure 1-15

The above figure is an overlay of two soundings performed 5 feet apart and one week apart. This sounding indicates a large amount of gassy liquid, which is encountered when refuse is saturated and is confined by fat clays used for daily cover. The landfill above is heavily irrigated to sustain lush vegetation.

25

The PPT investigation indicated that the irrigation water was percolating down and was being perched on daily cover layers or traffic decks. The water then migrated towards the interior of the landfill on these traffic decks and when it encountered the LFG collectors with their gravel packs it flowed downward to the bottom of the landfill.

As the water level increases and saturates the refuse more LFG is generated in the liquid, which creates gassy liquid. This process also causes rapid settlement, which can create excess pore pressures.

This is the greatest threat to slope stability in landfills as the refuse becomes saturated. If a seismic event should occur, the gas pressure will be quickly released and this could cause slope failure.

It also causes either flooded collectors and/or collapse of the collectors, which impedes the removal of the gas pressures. LFG pressures have been measured as high as 237 psi in these conditions. As the PPT instrument was removed from the landfill gas pressure was released in a roar and liquid foam bubbled up from the hole and it was effervescent. De-watering/LFG collectors have been installed in these areas and it has been very difficult to pump out the liquid due to the foaming condition. Centrifugal pumps do not work because the gas comes out of solution and creates only foam.

Bladder pumps must be used with very slow flow rates and the slow release of LFG out of solution makes gas production ineffective. The gas will not come out of solution unless it is agitated.

Please note above that the pore pressure was reduced by about 50% from the first sounding but the overall profile remained the same showing the repeatability of the instrument.

Chapter 2

This has been one of the most difficult changes in the industry I have been trying to sell in my career. It is like a religion to most people in the industry and telling them that smaller is better than bigger is like telling them there ain't no God. Hopefully the following chapter will create a few converts.

2.0 The History Of Landfill Gas Collectors

There have been many types of collectors installed in landfills since the regulatory agencies have required that LFG migration be controlled. When the environmental industry was being developed, the first environmental concern was with contaminated groundwater from leaking underground storage fuel tanks.

The environmental consulting industry developed a groundwater monitoring well design in order to obtain water samples for testing. This design involved drilling a bore hole down into the groundwater and installing a steel or PVC casing with perforated sections to allow water to enter. To prevent silt from plugging the well screen a sand pack was placed around the well screen to act as a filter.

The diameter of these wells began with a diameter of 2" and the wells were baled by hand prior to taking a sample. As the need for deeper wells and the agencies requirement of additional purging of the monitoring wells, it was necessary to increase the diameter of the wells to at least 4" diameter or larger to accommodate a submersible pump.

It is necessary to mention at this time, that the diameter of a well does not increase the flow rate of the medium being pumped whether it is liquid or LFG.

To increase the production of the well it is necessary to expose the screen section to more medium, by lengthening the screen section not the diameter of the well.

The flow of material (liquid or gas) into a well is controlled by the density or permeability of the surrounding soil or refuse and the availability of the material (gas generation) being pumped.

When the regulatory agencies required that LFG be removed from landfills and destroyed, the consulting industry just naturally applied water well technology to landfills.

Although flow mechanics can be applied to liquid and gas in many ways, unfortunately a landfill is a very dynamic environment and vacuum requires additional considerations. When a collector is installed with a sand or gravel pack it creates more problems than it solves. The sand or gravel pack around a collector pipe is more permeable than the refuse so vacuum will naturally flow upwards towards the surface rather than move through denser refuse. This causes short-circuiting and introduces air to the gas stream and reduces the effective vacuum influence of the collector.

For the past decade many approaches have been developed to install a LFG collector into a landfill. Most collectors were drilled in using a hollow stem auger an open flight auger or a bucket auger. Also, air and mud rotary, have been used. All of these methods produce drill cuttings, which have to be handled and disposed of.

Most of the collectors installed in the past contained large diameter pipes so that a submersible pump could be installed to pump out liquid, which may be flooding them. It is not always obvious that the drill rig is drilling through a layer of liquid, which can later impact the collector. There have been times when the drilling was suspended for the night at the mid-depth of the borehole, with no indication of liquid in the borehole. In the morning there could be several feet of water in the bottom of the borehole.

This condition is common due to the fact that most collectors are installed were LFG is being produced and this is caused by the presence of liquid. Most drilling methods disturb the trash prism to the point where detecting liquids is very difficult.

Therefore, blind drilling through the refuse and liquid layers and installing a large diameter pipe is not to carry the amount of LFG being produced but to accommodate a submersible pump.

Due to the way a landfill is constructed it is necessary to install many LFG collectors to obtain adequate vacuum coverage of the trash prism.

In most landfills refuse is compacted in an area of the landfill and is then covered by soil at the end of the day as per current regulations. This procedure creates contained cells that inhibit the flow of LFG and vacuum influence.

Since landfills are usually wider than they are higher it requires many vertical collectors to intersect as many cells as possible. It is not possible to intersect all of the cells by a collector but the lower the cost for each collector will allow more collectors to be install for better vacuum coverage.

Due to the restrictions of the cell construction each collector will influence only a certain amount of cubic feet of refuse and only a certain amount LFG will be produced within this zone of influence. It only makes sense to size the collectors to match the amount of LFG the collector will be exposed to and not just to allow the installation of a submersible pump. It is possible that a collector may only recover 50% of the LFG that is available around a given collector allowing 50% of the LFG to migrate offsite. PPT soundings have indicated vacuum layers and gas layers adjacent to vertical collectors.

2.1 Horizontal Collectors

More current landfills have begun to install horizontal collectors as the landfill is filled with refuse. One advantage to horizontal collectors is that more LFG is being recovered as the landfill is being filled. Vertical collectors are usually installed only after a portion of the landfill is filled, which could allow a substantial amount of LFG to escape from the landfill. One disadvantage of the horizontal collectors is that they may be isolated from layers of LFG caused by traffic decks and dense daily cover layers.

29

There may be a pancake effect of the vacuum influence with little vertical connection through the dense layers. Another disadvantage is that they may settle with the refuse and develop a dip in the pipeline and flood with condensate. At this time additional PPT Profiling is required to better understand the effectiveness of horizontal collectors. To expand the effectiveness of this method a combination of horizontal and vertical collectors may be required.

Traffic decks and dense daily cover layers can block the influence of vacuum on a horizontal plane. Shallow collectors that are installed near the top of the landfill are less effective because they short-circuit very easily; therefore, their zone of influence is limited. This should dictate the spacing of the collectors, as being closer together as they approach the top of the landfill, but this is not usually the case. Due to changes in the future plans for a landfill it often causes the top row of horizontal collectors to be too shallow when a future phase is canceled and these collectors are less effective.

In conclusion, it appears that it is more cost effective to install more, small diameter and cheaper collectors than fewer larger diameter and more costly collectors.

2.2 Push-in Gas Collectors

As stated above it may be more cost effective with improved vacuum coverage if more and smaller diameter collectors are installed instead of fewer large diameter collectors. The cone penetrometer truck can be used to push-in 2" diameter steel collectors at about a third of the cost of large diameter drilled in collectors.

Following a PPT sounding the PPT rig can be used to hydraulically push steel collector pipes into the waste prism at the best location for LFG extraction.

Various material types (such as stainless, galvanized, schedule 80 black steel) have been used in the past but according to machinists, stainless steel does not stay stainless outside the atmosphere and will corrode just as much as black steel. The oilfield industry uses thick wall pipe and allows the corrosion to create a patina layer of rust, which will limit the amount of decay of the pipe in harsh environments like oil deposits and landfills. Typically 5-foot lengths of mill slotted segments (which are also proven oilfield well design) with slot sizes (0.01" to 1/8") have been used. Therefore to save costs only schedule 80 black pipe is now used. The blank risers are also schedule 80 black pipe threaded and coupled. Flush thread pipe has been tried but the treads will not withstand a hard push if required and they can't withstand the tension if they have to be pulled up or out later. A typical 50-foot deep installation takes approximately 30 to 40 minutes to complete.

Figure 2-1

The main cause of well screen plugging is from liquid flowing into the collector, carrying silt and calcium carbonate, which blocks the sand/gravel pack and the well screen. Since push-in collectors are not installed in liquid zones using the PPT profiling method, it is not necessary to install a sand/gravel pack therefore there is no need for a 4" collector to accommodate a submersible pump. It is much more cost effective to install a 2" diameter steel pipe and well screen. The amount of LFG that will ever be produced in the area of the vacuum influence will never exceed the flow capacity of a 2" pipe.

Most lateral connections from a large diameter collector to the header are usually a 2" diameter pipe, which proves that this diameter can handle the flow rate. Also, the flow rate of this large diameter collector is controlled by a valve, which is rarely opened more than ¼" opening.

Why spend the money on drilling and installation of all of this large pipe and gravel to have a flow capacity though a ¼" opening in a valve.

If a 2" diameter collector is installed and if over time liquid should migrate and flood the collector, it is simple processes to hook a chain to the top of the steel collector and with a backhoe pull the collector up a few feet out of the liquid.

The collector is then reconnected at the surface. If no liquid is encountered at a location but multiple gas plumes are located at various depths, screened sections can be placed only at the gas producing zones reducing the possibility of air infiltration and liquid migration. This also reduces cost since slotted pipe costs three times more than blank riser pipe.

It is even more cost effective if the push-in wells are installed immediately following the PPT. The data being produced in real time can be used to design the collector immediately while the PPT rig is still over the location. Since a pilot hole has been provided by the PPT, the hole is expanded to a 3" diameter by a 3" diameter mandrel and a 2" collector can be installed very rapidly. This also saves the time spent grouting up the PPT hole and coming back later to pre-punch a pilot hole for the collector.

Again no cuttings are produced in the push-in method. This is a large saving in costs, especially in closed landfills where offsite disposal costs would be high.

Figure 2-2

The biggest objection to push-in wells is that it is believed that the screens will plug up with refuse. Opponents believe a gravel pack is required to prevent plugging of the screen. This is definitely the case with water wells however oil wells do not use a gravel pack but oilfield mill slots instead.

Many opponents have begun to use mill slots in plastic pipe but still drill and use a gravel pack. It is generally accepted that a good producing well with a gravel pack will plug up faster (and they do plug up) than a mediocre producing well due to the increased amount of particular material being moved through the gravel filter pack.

The same is true with push-in wells the better producer will plug before a mediocre well. However, another advantage of the push-in wells is that they can be cleared by injecting high-pressure steam down the well and through the screens. A drilled in plastic well cannot be cleared. The use of these procedures will drastically reduce the operating and maintenance costs of landfills and assist in maintaining regulatory compliance.

Advantages of the PPT method over conventional drilling methods are as follows:

Conventional Method	PPT Method
• Boring log based on one person	• PPT log can be intrep. by many
• No gas profile prior installation	• Gas verified prior to well install
• No liquid profile prior to install	• Liquid zones can be avoided
• Depth control is marginal	• Depth control is exact
• Inclin. of bore hole unknown	• Inclin. recorded by inclinometers
• Max. Health & Safety concerns	• Min. Health & Safety concerns
• High well material costs	• Low well material cost
• Must dispose of cuttings	• No cuttings disposal
• Flooded well must be pumped	• Well can be pulled out of leachate
• Must dispose of leachate	• No leachate disposal
• Cannot be reused	• Well can be removed and reused
• Subject to melting and crushing	• Steel will hold up better
• Cannot inject steam	• Can be used as a steam injector
• Cannot be unplugged by steam	• Can be unplugged by steam
• Slow installation time	• Very fast installation time

Many consultants will claim that they have tried the push-in collectors but they did not work. When I asked them if they used the PPT first to verify if there was gas at the location, they would say no. I would then ask if they had every drilled in a large diameter collector and had very little gas recovery and they would say yes. Then I ask why you reject the push-in technology and not the drilled in method when you probably installed both collectors in an area where there is no gas.

After all this money is spent on a large diameter well, the valve at the top of the collector is usually opened only about a ¼ of inch, it is not cost effective to pay for a large diameter collector and only use a fraction of its flow capacity. The 2" diameter collector can have its valve opened for full flow without short-circuiting unless it is very close to a slope.

Assuming the collector is not close to a slope, it is usually found that the depth of the collector controls the radius of vacuum influence. A 100-foot deep large diameter collector will have a zone of influence of about 90 to 95-foot radius. To try and increase this radius by opening the control valve will cause short-circuiting and air will enter the trash prism. For simplicity (Figure 2-3) shows the amount of LFG a large and small diameter collector will be able to extract. In actuality vacuum fans out in layers like fingers in all direction. The PPT has encountered vacuum layers and gas pressure layers in the same sounding.

As indicated in the Figure 2-3, the small diameter collector will extract the same amount of LFG but at one-third the cost. The large diameter collector could cost three times as much as a small diameter collector but will never extract 3 times the amount of LFG.

Large Diameter Collector   Small Diameter Collector

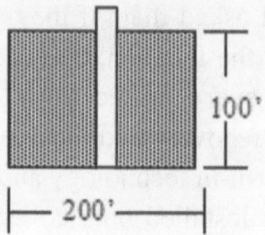   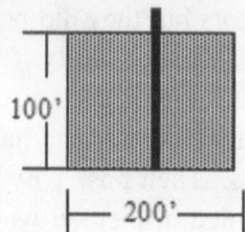

Figure 2-3

The other controlling factor is the permeability of the trash and the amount of gas being generated. Once the gas is pulled out of the gravel pack of a 24"- 36" diameter bore hole the recovery of the gas into the gravel is dependent upon the permeability of the refuse around it and the amount of gas being produced.

Until the oilfield mill slot was used, a popular large diameter screen design and still in use today, is to drill four ½ inch holes 90 degrees from each other every 3 to 6 inches along a 4" or 6" diameter pipe. This large diameter pipe does have a higher flow capacity than a 2" diameter pipe but the flow capacity is restricted by the few ½ inch holes drilled into it.

To compare the flow restrictions of the two types of screens the following calculations were performed. The mill slots of the 2" diameter screens are 1/16" wide and 3" to 3.5" long with 12 slots per row. This calculates out to 2.28 sq. in. of open space per row.

The area of the open end of the schedule 80, 2" diameter pipe, (which is the lowest restriction possible of any size pipe) is 3.1416 sq. in. Therefore, just 2 rows of slots (i.e. 6" of screen) provides more area of open space (4.56 sq. in.) than an open ended 2" diameter pipe.

According to the specifications for the 4" collectors, four ½" holes are to be drilled at 90 degree angles and 3" apart. This provides only 2.36 sq. in. of open space for every 6" of screen. It will take 31.8 inches of screen to equal the open end of a 4" collector.

Area of a slots = .19 sq. in. x 12 = 2.28 sq. in. per row

Area of ½" hole = .19635 sq. in. x 4 = .7854 sq. in. per row

Mill slots provide 3 times less restriction than the current large diameter pipe collectors do.

Another factor is time. With additional holes in the larger collector, LFG can be removed faster with a 4" diameter pipe but will have to be throttled back in a short time to prevent overdrawing the landfill. It is like placing an oversized pump in a water-well.

A 10 gallon per minute (gpm) pump in a well with only a 5 gpm recovery rate will cost more to buy and operate and be off half the time. It would be more cost effective to buy a 5 gpm pump. The same logic should be used for LFG collectors.

In conclusion, the cost of the large diameter collectors is 3 to 4 times higher than the 2" diameter collector, but they will never recover 3 to 4 times more LFG out of the larger collector.

Chapter 3

Millions of dollars are spent every year installing LFG collectors into our landfills across the country and I bet we are not collecting even 50% of the available gas that is being generated. Read on to find out why.

3.0 Vacuum Evaluation

It is often necessary to evaluate the vacuum coverage of a LFG collection system due to high readings from perimeter probes.

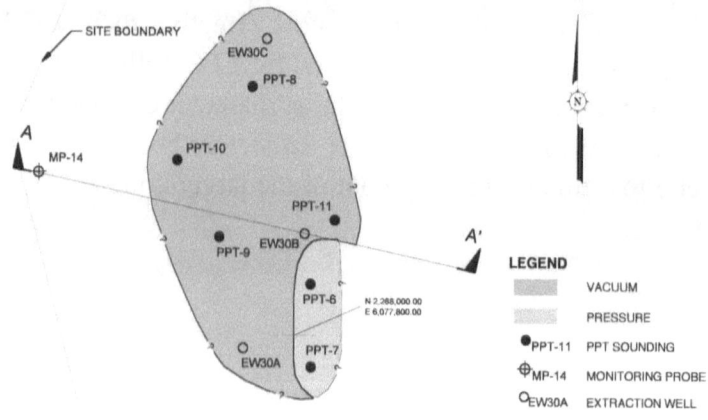

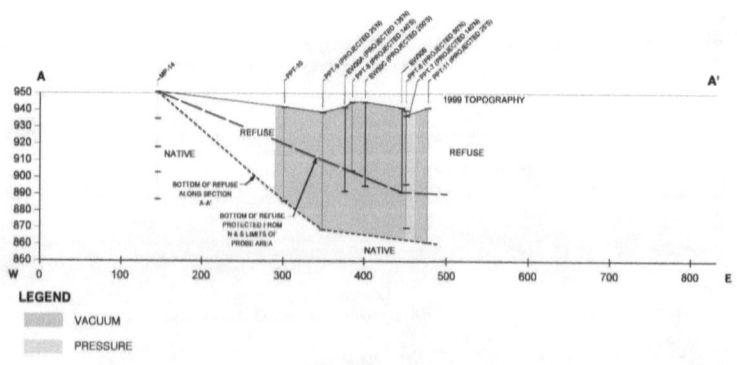

Figure 3-1

38

As the Figures 3-1 indicates the collectors in this area are providing good vacuum coverage except near EW30A & EW30B where LFG pressure was indicated in PPT-6 and PPT-7. The purpose of this investigation was to determine why MP-14 perimeter probe was showing an increase in methane levels.

It was later determined that the vacuum coverage was adequate when the vacuum system was operating. However, the valves on the collectors were not turned off when the power plant was shut down so the header became pressurized from high producing collectors so the above collectors became injectors and impacted the perimeter probes. Back flow prevention will be discussed later in this book.

3.1 Vacuum Influence Testing

In April 2003, STI Engineering installed 31 landfill gas (LFG) collectors into a 28 acre closed landfill in Southern California. The collectors were push-in 2" diameter steel schedule 80 pipe with 1/8"x 3.5" long mill slot screens. The depths ranged from 20 to 40 feet below ground surface. To verify the spacing of the collectors an Influence Test was conducted on one of the collectors. Time did not allow for more collectors to be tested.

Pre-Qualifying Locations

It is STI's approach to pre-qualify the selected locations with the use of the Piezo-Penetrometer Test (PPT) prior to installing a push-in collector.

The PPT data will verify any gas pressures that may be present, also identify any liquid layers that should be avoided to prevent flooded collectors or providing a conduit for liquids to the bottom of the landfill. Also the gas collection system will be more effective if the vacuum influence from the collectors do not overlap too much.

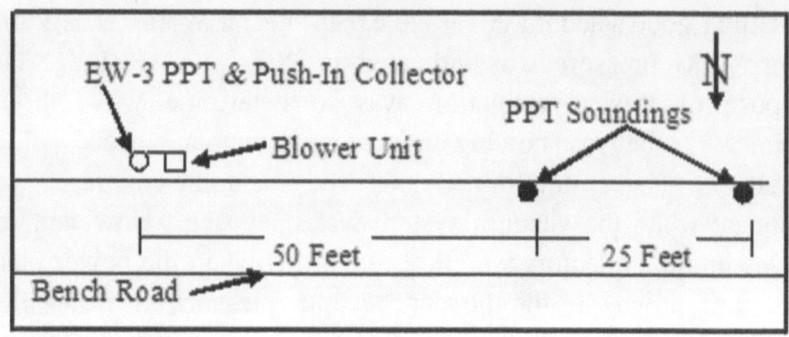

Figure 3-2

In this figure, refuse is indicated from 5 to 30 feet below ground surface with intermitted layers of gas pressure and some larger spikes from moist refuse indicated at 5 to 25 feet. The negative pressure readings in the log below are caused by the suction pressures on the PPT transducer, as it passes through dry soil and refuse and where the gas pressure is not high enough to overcome this effect.

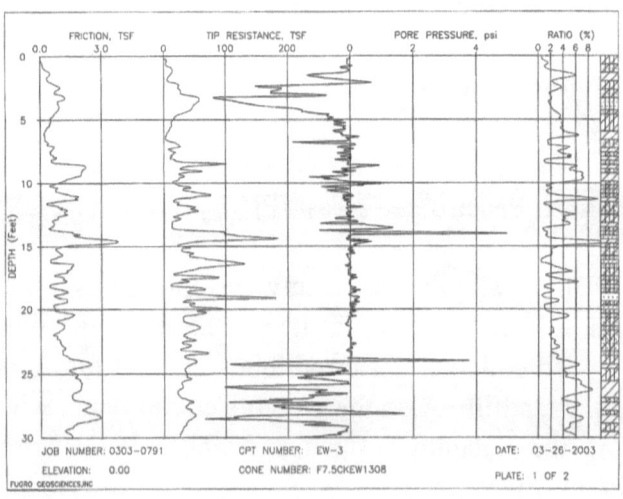

Figure 3-3

40

A 2" diameter steel push-in collector was installed at this location. The screen sections were placed at 10 to 30 feet below ground surface with 10 feet of blank riser.

3.2 Influence Testing

To assist in determining the spacing of the collectors it was decided to perform an Influence Test on one of the collectors. Following the collector installation, a 1.5 hp blower was connected to the collector. The blower had a flow rate of 75 scfm and a capacity to produce 18 inches of water column vacuum. During the test, the vacuum gauge read 1.5 inches of vacuum indicating little restriction and full flow from the refuse. The blower ran for about 30 minutes to remove any excess gas pressures in the zone of influence, as the PPT rig was moved 50 feet west from the collector where another PPT sounding was performed.

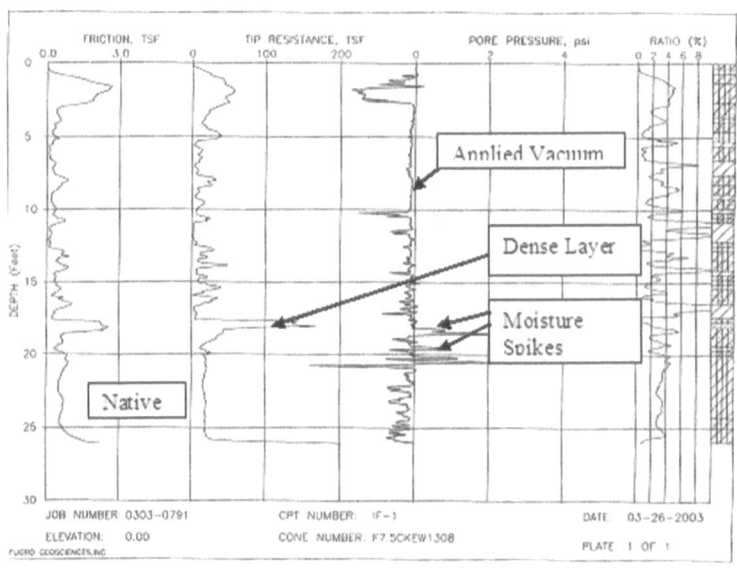

Figure 3-4

41

As indicated in the pore-pressure column of Figure 3-4, the blower had applied a vacuum from 3 to 26 feet below ground surface.

At IF-2 the positive pressures indicated, at 17 to 21 feet bgs are in moist layers of refuse where the vacuum cannot transmit through. As pointed out in figure 3-5, the positive pressure is created as the cone passes through the moist dense layers and liquid is forced into the transducer chamber. The log also indicates that the native soil was encountered at 20 feet below ground surface. This log indicates that the vacuum is transmitted through the native soil verifying the permeability of the native soil and why the LFG has migrated to the groundwater at this site.

Please note the tip resistance of the cover soil is about 50 tsf and that the native soil has a tip resistance of only about 20 tsf. This is another reason that LFG migrated to the groundwater.

Once the above sounding was completed, the PPT rig was moved another 25 feet west from this sounding location, which was 75 feet west of the collector where another sounding was performed (see Figure 3-5).

As indicated in Figure 3-5, the vacuum influence was apparent in the shallower depths but pressures were indicated in the more moist refuse starting at 17 feet bgs. Refuse was indicated at 3 to 18 feet below ground surface, where native soil was indicated. The blower had been operating approximately 1.5 hours at the conclusion of the second PPT sounding.

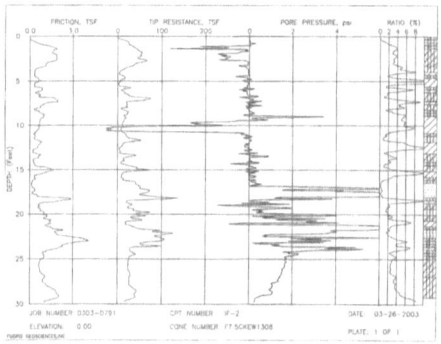

Figure 3-5

The methane concentrations were not measured during this test, as it was not considered an accurate measurement due to the recent installation of the collector.

3.3 Conclusions

The following are some conclusions based on the findings of the Vacuum Influence Tests performed on this small diameter push-in collector.

- The PPT detects the effects of vacuum on gas generating refuse.
- The effects of the vacuum occurred soon after it was applied.
- The vacuum influence exceeded the distance of the depth of push-in collector by a factor of over 2 times. Example: This collector was 30 feet deep but the vacuum influence was over 75 feet of radius. This rarely occurs with drilled in collectors where short-circuiting is common. A 30 foot drilled in collector would have an influence radius of about 25 to 30 feet before short-circuiting would occur.
- Moisture conditions of the refuse has an impact on the vacuum influence.

43

- The mill slot screens allowed full flow from the blower.
- Vacuum influence appeared to diminish with depth. This could be due to moisture content, density, (hence permeability) of the refuse or the size of the blower and time.
- No short-circuiting was visually apparent around the base of the collector although no gas analysis was performed to verify this.

It would have been interesting to perform additional tests on the bench road below this location to see if the influence moved downward through the refuse, but time did not allow this.

Future Influence Studies should also include performing PPT soundings in all directions and at greater distances. This same test should be performed at landfills that have horizontal collectors to determine their zone of vacuum influences as well. Just prior to performing this test at this landfill, STI had performed 13 PPT soundings at another nearby landfill with horizontal collectors.

Some of these soundings were performed in the center of the landfill to full depth and some of them were around the perimeter of the landfill. None of the PPT soundings indicated any vacuum in these areas. Eighteen vertical collectors were soon installed into the landfill.

Most landfill operators assume their collection system is adequate until grid walks, perimeter probes or groundwater-monitoring wells indicate a problem. By performing Influence Testing and using the PPT, a collection system can be better designed, installed and maintained.

Some influence tests have been performed by other consultants on drilled in collectors using drilled in probes with screen sections at various depths. The problem with this approach is the drilling process disturbs the waste prism to the point as to change the permeability of the area being tested. The screen section can be so long as to connect areas in the landfill with vacuum layers that normally would not have occurred if the probe were not installed.

Learning of this condition by comparing results with a PPT sounding did enable STI to develop the Internal Conduit concept.

3.4 Internal Conduits

It is not uncommon for PPT soundings performed in waste prisms with an existing LFG collection system to encounter an applied vacuum from a nearby existing collector and gas pressures layers in the same sounding. As expected due to variations in density and permeability, the zone of vacuum influence around a typical LFG collector in MSW looks like a series of fingers rather than a uniform sphere.

When this condition is encountered it is not always necessary to install another collector to improve the vacuum influence of a collector.

Instead a 3/4" perforated PVC pipe can be installed using push-in methods, which connects the vacuum zone(s) to the gas pressure zone(s). A GMF flush joint casing with a dummy tip is pushed to the target depth down a previous PPT hole. A 3/4" perforated PVC pipe is inserted down the GMF casing. The GMF casing is then retracted leaving the dummy tip and Internal Conduit in place. The gas pressure now has a pathway to the vacuum zone and to the collector. Once the conduit is installed at the correct depth, a washer type plug is placed on top of the PVC pipe and the top of the PPT hole is grouted up. This procedure is most effective along the perimeter of the landfill where small wedges of refuse are encountered and are often the cause of surface emissions.

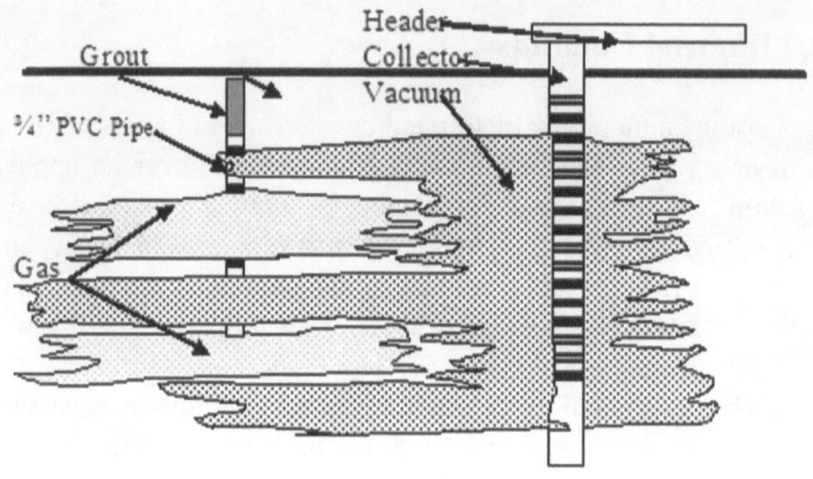

Figure 3-6

3.5 Cluster Collector Design

The following is an alternative to conventional vertical and horizontal collector systems by combining both approaches. This approach has not been used at a landfill at the time of this publication but it is planned in the near future.

Many landfill operators have tried to keep their landfills in compliance by recovering the LFG being generated by the fresh refuse as soon as possible. Currently, this is accomplished by installing horizontal HDPE pipes wrapped in gravel and sometimes in geotextile fabric as the refuse is being placed in the landfill. This has proven to be an effective approach in most cases until the refuse begins to settle and the horizontal collectors begin to bow and fill with condensate. If the collector is not connected at both ends to the vacuum system then the collector will loose most of it's effectiveness. The pipes are also prone to crushing as the refuse settles.

Vertical or horizontal collectors have advantages and disadvantages so it would appear that to have the best of both worlds would be to take the advantages of both and combined them into the most effective system.

Some of the disadvantages of vertical collectors is the possibility of short circuiting to the surface and the cost of connecting each collector to the header with the cost of monitoring and adjusting each valve.

A horizontal collector is like installing a header inside the landfill however these collectors can create a pancake vacuum layer with limited influence on the vertical plane depending on how the landfill is constructed. To increase the vertical influence from these horizontal layers it is necessary to penetrate the overlying dense layers by installing vertical collectors.

In the past the PPT has indicated vacuum layers running hundreds of feet through a landfill at a given elevation through a porous layer. The PPT has also indicated layers of gas pressure above and below this vacuum layer. As outline in the section about Internal Conduits these layers can be connected.

To create a vacuum layer in refuse already in place a vertical collector is pushed-in with the screen sections in a low density and porous layer detected by the PPT (for example at 40 feet below the top of the landfill). A vacuum is applied to the collector and the PPT is used to track the vacuum influence from the collector.

At approximately 100 to 200 feet away from this collector another collector 2" diameter is pushed into the landfill with screen sections intersecting the vacuum layer at 40 feet below the surface of the landfill. The collector is screened to within 15 feet of the surface and is then capped and the hole above the screen sections is sealed with grout. (Let's call these collectors inverted collectors.) This procedure is repeated in all four directions providing a large area with vacuum influence but with only one valve and header connection.

To create a more uniform vacuum connection a porous vacuum layer can be installed as the landfill is being constructed.

At pre-selected elevations (such as 40 feet intervals) a traffic deck is created. Instead of installing a horizontal collector with the cost of HDPE pipe a windrow of gravel or shredded tires are placed in a grid pattern across the traffic deck as the refuse is placed and is surveyed in to record its location for future reference. Once a 40 foot thick layer of refuse is placed, the vacuum collector is installed at the location that will intersect the gravel/tire layer. The PPT is used to verify the vacuum through the gravel/tire layers and the inverted collectors are installed.

Imagine controlling 5 collectors with only one header connection and one valve. For monitoring purposes a 1/2" tube could be connected to the top of the inverted collectors and brought to the surface and capped. This would be used to verify the vacuum effectiveness of the system over time.

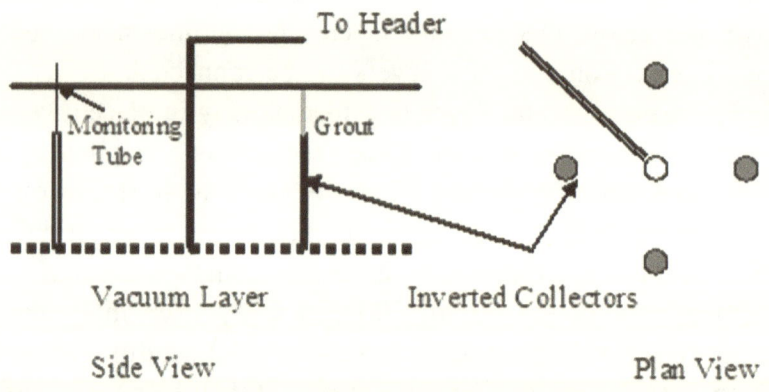

Figure 3-7

Center collector has screens only on the lower half, for higher vacuum capability. If the landfill is being constructed in phases such as 40 foot thick cells another gravel/tire windrow would be placed in the four directions of the inverted collectors.

48

The center collector would be reinstalled after the next phase is placed over the first center collector. The depth of the collector would overlap the previous collector to the top of the bottom collector's screens. Four more inverted collectors would be installed and their depths would overlap the first inverted collectors to continue the vacuum coverage downward and laterally.

As the cost of pipe and other O&M costs continue to rise, this alternative may be necessary to reduce gas extraction costs.

Chapter 4

It is amazing to me that a monitoring device as important as detecting migrating methane gas is found to be defective and the governing bodies that mandate that they be installed fail to take action. It is also amazing that there are thousands of professionals in the industry that continue to install Multi-Chamber Probes without question.

4.0 Multiple Chamber Probes

Current state and federal regulations require that probes be installed around the perimeter of active landfills to monitor migration of landfill gas (LFG) through the surrounding native soils. It is also required that inactive or closed landfills that indicate a potential risk to the environment or human health must have perimeter probes installed and monitored. There are no current regulations requiring the probes to be tested for integrity following installation or after many years of operation.

4.1 Probe Chamber Location vs. Soil Profile

To save costs, multi-chamber probes are permitted (not required as some people believe) to be installed in the same borehole instead of single chamber probes at different depths in separate boreholes.

According to California Integrated Waste Management Board (CIWMB) regulations, it is recommended that the chambers be installed at 3 or 4 different depths. One chamber should be 10 to 20 feet below ground surface (bgs) one at the center depth of the thickness of the refuse, another at the bottom depth of the refuse and sometimes, one 25 feet below the bottom of the refuse, if the groundwater level will allow it.

It is assumed that the chambers will be installed in the more porous soil layers but discreet soil sampling to locate these layers is not specifically required in the regulations.

50

The common practice is to log the cuttings from the drilling process, which is not very accurate at determining the depth of thin sand or clay layers that can affect LFG migration. The intent of perimeter monitoring would be much better satisfied if discreet sampling was required in non-uniform soil profiles.

Unfortunately, due to the contract award being typically based on low bid, if a consultant proposes that he will take discreet samples to confirm a porous layer the bid will be too high and he will not be awarded the contract. Some thought should be given to the need of multi-chamber probes in a soil formation that has the same porosity. If the entire soil profile is highly porous the LFG can easily move from one chamber level to the next through the surrounding soil particularly when vacuum is applied to evacuate the probe chamber.

4.2 Seals Between Chambers – Integrity Test

The diagrams provided on the CIWMB's web site indicate that a 5-foot thick Bentonite seal should be installed between the chambers. Figure 4-1 presents a schematic of a typical multi-chamber probe. However, there is no requirement to test the chambers for seal tightness following construction. Also, there is no regulation requiring that the probes be tested over time to insure the Bentonite seals have not dried out and cracked, causing leaks.

To determine the condition of seals in multi-chamber probes the author developed an Integrity Test, which verifies if the chambers are isolated or if they are leaking and cross connected.

The following are the primary tasks involved in an Integrity Test:

- The boring log for the multi-chamber probe should be reviewed prior to the field test to verify the type or types of soil in which each chamber was installed. (With no discreet sampling, it is better than nothing.)
- Pressure/vacuum readings are taken and recorded from each chamber.

- Methane readings are taken and recorded from each chamber.
- Nitrogen pressure is applied to the deepest chamber.
- A very sensitive Magnehelic gauge is successively connected to each chamber as the nitrogen pressure is applied.
- If no change in pressure is detected in the chamber above the bottom chamber, then the pressure tube should be attached to the next upper chamber. This process is repeated all the way up to the top chamber.
- If there is no change in pressure in any of the chambers then the Bentonite seals are tight.
- If pressure is indicated in a chamber and the response time in a chamber is fast, then the Bentonite seals are cracked and leaking.
- If the response time for the pressure change in a chamber is slow then the pressure may be leaking through the surrounding soil. This determination should be made based on the review of the soil boring. If the soil boring indicated that the surrounding soil is coarse sand or gravelly, a multi-chamber probe should not have been installed in this location.

While performing Integrity Tests on three chamber probes in several counties in California it was discovered that 3 out of 5 probes failed the test. From this discovery the author has taken a closer look at the overall installation and monitoring of multi-chamber probes and the regulations governing them.

During an Integrity Test at one of the landfills in Southern California the reaction time for the nitrogen pressure to travel from the bottom chamber to the top chamber was so fast that the only conclusion was that the Bentonite seals had desiccated, cracked and were leaking. Had there been a longer lag time between the nitrogen pressure from the bottom chamber to the top chamber, then it would have indicated that the nitrogen was flowing through the soil formation.

4.3 Potential Impacts of Leaking Probe Chambers

It is now generally understood in the industry that most groundwater impacts from landfills are caused by landfill gas migration. Usually, the Regional Water Quality Control Board (RWQCB) only reviews landfill perimeter probe records after a groundwater monitoring well has indicated that the groundwater has been impacted. This is done to help in determining how and where the groundwater is being impacted. Unfortunately by the time contamination is detected in the groundwater wells, it is likely too late for intervention. It is time for mitigation or remediation.

Perimeter probes are primary indicators of the effectiveness of the gas collection system in a landfill and can provide early warning of potential impacts to groundwater if properly constructed, tested and monitored. Currently, the Law Enforcement Agencies (LEA) do not require action until the perimeter probes indicate a methane level of 5%, i.e. the lower explosive limit of methane. At 4% concentration, impacts to groundwater quality can occur should the LFG (which contains many toxic constituents) come in contact with the groundwater.

Also, the methane levels could be much higher if the probe chamber seals are leaking.

If most of the probes are indicating that LFG is migrating laterally from a landfill, whether or not the measured concentration have reached 5%, this can be an indication that the collection system is not adequate enough to prevent the LFG from impacting the groundwater.

Once a methane concentration greater than 5% is detected in a perimeter probe, it is usually necessary to design and install a LFG control system or investigate why the existing LFG collection system is not controlling LFG migration. To assist in design and investigation of LFG control, a method has been developed using the digital data obtained from the Piezo-Penetrometer Test (PPT) to create a 3-D Profile of the conditions inside a landfill.

This process is used to locate possible pathways of migrating landfill gas, which have impacted perimeter probes around a landfill. The method has been used at several landfills to investigate the cause of the migrating LFG into multi-chamber probes.

If the PPT method is to be used to investigate an impacted probe, it is necessary to verify the readings in the perimeter probe of concern. It has been observed that many multi-chamber probes have LFG in more than one chamber. It is important to know if LFG is migrating at two different pathways or if it is only one pathway and the LFG is leaking from one chamber to another. This will indicate whether to look for one source or two and to locate and design LFG collectors.

4.4 Flawed Monitoring Procedures

The leaky chamber conditions described above are compounded by the monitoring procedures. The current monitoring procedures do not include methods to check the integrity of the chambers.

It is assumed that once the probes are installed they will function properly, indefinitely or until obvious visual damage occurs to the top portion of the probe.

Let's review what is happening during the process of taking readings of a 4-chamber probe with leaking chambers. For this discussion we will assume there is substantial LFG in the subsurface at the depth of the third chamber. The technician attaches a tube from the gas analyzer to the shallowest chamber or the deepest chamber depending on the order in which, the technician is taking the readings   (He should take a pressure/vacuum reading first but many do not). If it is the shallowest chamber, the technician will run the vacuum pump long enough to evacuate at least 2 calculated volumes of the chamber. This chamber may be cross-connected to the next chamber or the surface either due to leaking seals or very porous surrounding soils.

If it is connected to the surface this chamber will always read low, if it is cross-connected to the next deepest chamber the vacuum applied to the first chamber has started the evacuation of the second chamber (Figure 4-1).

When the technician starts to evacuate the second chamber and if it is cross-connected to the third chamber, LFG will start to enter the second chamber. When the technician detects the increasing level of methane he will usually keep the pump running until the methane reading stabilizes. At this time, if air is leaking from the first chamber into the second and LFG is entering from the third chamber dilution will probably keep this chamber from reaching the 5% action level.

During this time LFG has been pulled out of the third chamber and could lower the concentration of LFG in chamber three below the action level especially if air is entering from the fourth chamber.

When the technician evacuates the third chamber the process will pull the LFG/air mixture from the second chamber and, if the third chamber is cross-connected to the fourth chamber it will be pulling air from that chamber lowering the LFG concentration in the third chamber.

This chamber will probably read below action levels also. When the technician evacuates the fourth chamber the process will pull LFG from the third chamber again, lowering the concentration even more in the third chamber.

It should be noted that the above conditions and the current monitoring procedures will always cause the probe to give a false low methane reading, not a false high reading. Of course if a leaky probe gave a false high reading the landfill owners/operators would be screaming to change the regulations.

The LFG that is measured in the chambers that drew LFG into them from the impacted chamber are giving a false indication that LFG is migrating from more than one layer from the landfill. Some landfill operators claim that it doesn't matter because the whole probe will just equalize at the same methane level.

This would be true if only one chamber was constantly read and there was no leak to the atmosphere and the top chamber.

However, this is never the case, all chambers are monitored and it is this procedure that causes the many exchanges within the chambers and dilutes the concentrations.

At this time it is not known how many multi-chamber probes may have leaking chambers. Based on reviewing the monitoring records of several landfills, the records seem to indicate that if LFG is detected in two or more chambers there is a good possibility that they are cross-connected. However, it can also indicate that LFG is migrating at two different elevations. Only an Integrity Test can verify which condition it is.

Also, an Integrity Test combined with a review of the boring logs of the probe can determine if the connection between the chambers is through the Bentonite seal or if the LFG is passing through the surrounding soil. A review of the boring log of the probe may indicate that the soil type is the same granular formation down the full length of the probe and therefore a multi-chamber probe should not have been installed in the first place. Only single chamber probes can eliminate the possibility of cross-connection.

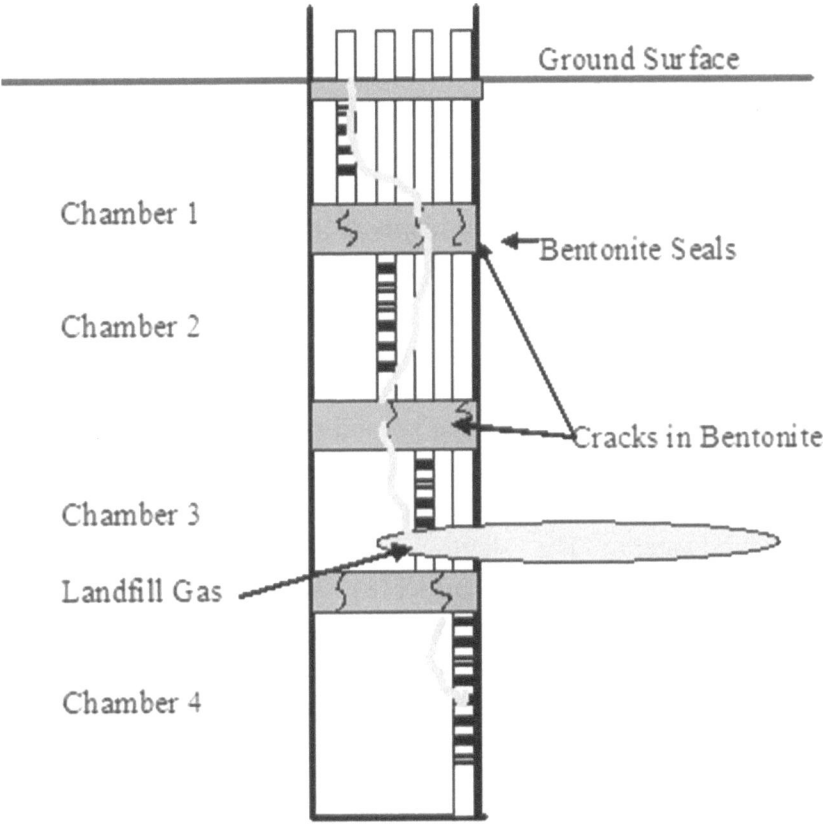

Ground Surface

Chamber 1

Bentonite Seals

Chamber 2

Cracks in Bentonite

Chamber 3

Landfill Gas

Chamber 4

Figure 4-1

A review of probe readings from a particular landfill can indicate the possibility that the chambers are cross-connected. If the readings indicate methane in two or more chambers no matter the concentration level then the "possibility" of cross-connection is indicated. An Integrity Test should be performed on these probes. However, if LFG is indicated in only one chamber and the other chambers are not under a vacuum influence then this can be an indication that the chambers are not cross-connected.

4.5 Suggested New Probe Approach

If the regulations were modified to require that future multi-chamber probes must pass an Integrity Test or be replaced, the landfill operators would most likely elect to install single chamber probes and thus avoid the added cost of testing and replacement if failure occurs.

Currently, three to four chambers are required in a probe to monitor different levels (Figure 4-1). If you consider the purpose of the perimeter probes, only two probes at two different elevations are required (Figure 4-2).

The slotted pipe of the shallow probe should be from 10 feet bgs to mid depth of trash. The second probe at least 5 feet away from the first probe would start with slotted pipe at just above mid depth of trash to just below bottom of trash depending on the depth to groundwater. The shallow probe will monitor the LFG migration that would cause surface emissions and LFG that would impact structures. The deep probe would monitor the migration of LFG that would impact groundwater. The 5 to 10 foot overlap at the mid-trash level would assist in determining if LFG is migrating from the mid-range of the landfill by indicating LFG in both probes. This could also indicate LFG is migrating from the full depth of the refuse. The use of two separate probe chambers also eliminates the importance of identifying different soil layers during drilling. Cross migration through various soil layers will not affect the readings from each individual probe.

As stated earlier, water quality boards often use the perimeter probes to indicate where the groundwater impacts may be coming from. However, the location of the probe is important if it is going to be of any assistance in this type of investigation. The regulations are not too specific as to the location of probes as long as they are within the property line of the landfill.

This can allow a perimeter probe to be installed hundreds of feet away from the limits of the refuse.

The groundwater can often be only tens of feet from the bottom of the refuse therefore; the migrating LFG can be in contact with the groundwater long before it reaches a perimeter probe. Perimeter probes should be installed such that the horizontal distance of the probe from limits of refuse is equal to or less than the vertical distance of the bottom of refuse above the groundwater level.

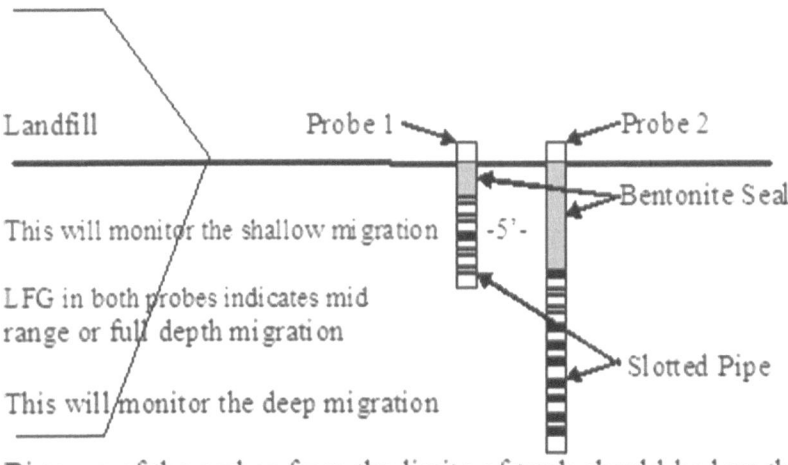

Distance of the probes from the limits of trash should be less the distance of the groundwater from the trash bottom.

Groundwater   Not To Scale

Figure 4-2

Since LFG usually migrates faster along the strata of the soil, than going across it, this should provide some warning before the groundwater is impacted.

59

4.6 Vacuum Influence on Probes

Currently, landfill operators have to monitor their perimeter probes at least once a month. The CIWMB regulations state that a probe should <u>not</u> be in the vacuum influence of a nearby collector but in reviewing monitoring records at several landfills many of the forms do not have a column for recording any pressure or vacuum readings. It could be assumed that these readings were not taken during the monitoring procedure.

In some cases it has been observed that collectors have been installed in native soil outside the trash prism and adjacent to "hot" probes to pull the methane gas out of the impacted probes. These probes with vacuum readings should be disqualified because they will not be able to monitor the migration of LFG outside the vacuum influence zone.

A probe in the influence of a collector can be "blind" to LFG migrating only 10 feet away from the probe. If the records indicate a probe that is close to the limits of trash and yet a zero methane reading is recorded in all the chambers, then it may be in the vacuum influence of a nearby collector and the probe should be tested for vacuum influence. A probe that is detected to be under the influence of a vacuum should have the vacuum reduced or another probe installed adjacent to it, but outside the zone of vacuum influence. Perimeter probes should monitor the static soil conditions surrounding a landfill.

4.7 Back Flow Prevention

PPT Profile investigations have identified another possible cause of impacted probes, groundwater impacts, and surface emissions.

Typically, collectors that are installed on the perimeter of landfills to control LFG migration are not high producing gas collectors. Their main function is to collect low quality gas that could otherwise migrate, offsite.

However, when the flare or power plant is shut down, the header can become pressurized from the high-producing collectors in the center of the landfill. The LFG can then back flow into the low producing perimeter collectors allowing the LFG to migrate to the probe and possibly to the groundwater.

This is another reason why perimeter probes should not be in the zone of vacuum influence of collectors. When you consider that the LFG pressure in the header could increase to 50+ psi and when the vacuum is re-established the LFG will be extracted at less than 1 psi. This indicates that much of the LFG that back-flowed into the surrounding formation would not be recovered.

One way to address this problem is to turn off the valves to the perimeter collectors prior to shutting down the vacuum system. This works well for planned shut downs but sometimes the vacuum systems go down in the middle of the night and it is difficult to close valves on a dark landfill. There is also the time consuming problem of resetting the valves when the system is back on line.

A better alternative is to install check valves on all perimeter collectors (Figure 4-3) which will prevent gas from entering the collectors, impacting probes or the groundwater, or causing surface emissions during vacuum system shutdowns.

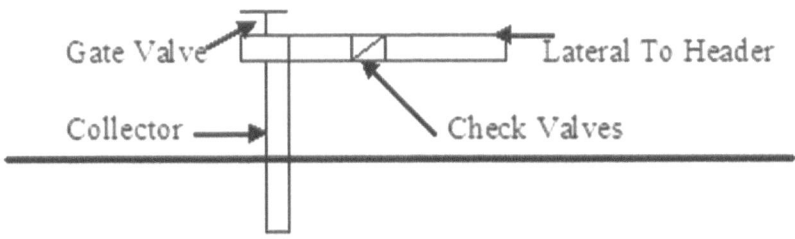

Figure 4-3

4.8 Pre-Qualify Probe Locations

The PPT can also be used to qualify a potential probe location prior to installing new probes.

Once a preliminary location has been selected for a probe, a PPT sounding may have to be performed inside the landfill to determine the depth of trash. If the depth is known, then only the probe location would be tested. The PPT will be able to identify the various soil types, their "relative" density and locate the more porous layers that will transmit LFG faster.

If the pore pressure transducer detects vacuum, gas pressure or a perched water layer in this location, consideration should be given to changing the location, thereby saving the cost of installing a probe in the wrong location.

Following a PPT sounding and if the location is found to be clear of vacuum, gas pressure and perched water, the PPT rig can push in slotted steel pipe or install PVC pipe to the determined depths to construct the probe. The PPT can install these probes at a fraction of the time and cost of drilling and no cuttings are produced. Many landfills are surrounded by alluvial deposits, weathered bedrock or fractured bedrock.

The PPT rig can apply 20 tons of force on a 15 square centimeter tip and can penetrate most alluvial deposits and weathered bedrock. If fractured bedrock is encountered, the chances of intersecting a fracture that can transmit LFG even using drilling techniques are very remote.

The fracturing condition in the bedrock should be considered in the decision tree, as to the effectiveness of placing a probe in this type of location. The PPT will locate the most practical locations for perimeter probes.

4.9 Conclusion

The author has discussed the situation of cross-connected probes with many California county waste departments, health departments (LEAs) and landfill operators.

As it stands cross-connected probes are in compliance with current regulations but are providing false readings of existing conditions. The author feels it is important to act on this situation before a major catastrophe occurs such as a structure or home exploding. Also, by placing reliable probes within the appropriate distance from the landfill in relation to the depth of groundwater, the environment will be better protected.

The following are suggested tasks and modifications to the regulations, perimeter probe construction and proper monitoring procedures.

- Audit current probe reading logs in each state by each county.
- Identify the number of multi-chamber probes that exist.
- Identify how many of these probes have LFG in two or more chambers.
- Identify how many probes in close proximity to the landfill may be under a vacuum influence by having a zero methane reading.
- Verify the readings on the above-identified probes with an Integrity Test.
- Identify how many probes are further away from the landfill than the groundwater.
- Once the audit and field verification is complete, this data should be used to modify the regulations to address the findings of the audit.
- The following modifications to the current regulations should be considered if the audit warrants.
- Disqualify any probe under the influence of a vacuum.
- Eliminate the use of new multi-chamber probes in a single borehole.
- Two separate chambers (shallow/deep) should replace existing probes that fail an Integrity Test.
- Existing multi-chamber probes should be Integrity Tested at least once a year.

- Perimeter probes should be installed no farther away from the landfill than the distance between the bottom of the refuse and the groundwater.
- Instead of using an arbitrary 1000 foot spacing, more probes should be installed in low density or granular soil and fewer probes should be installed in dense soil and fractured bedrock.
- Back flow/check valves should be required on all perimeter collectors. If the modified regulations require that the probes be closer to the landfill due to the proximity of the groundwater this becomes even more important.
- Probe locations should be pre-qualified using the PPT method.

Due to the probe construction and monitoring shortcomings discussed above it is possible that there are many more landfills with migrating LFG and at higher concentrations than what are being presently reported.

Once the regulations are modified and with the use of the Integrity Test and PPT technology with its in-situ capabilities, landfill engineering and compliance can be significantly improved.

Chapter 5

This is a case of two government agencies working at cross-purposes when it comes to small inactive landfills. The water boards mandate that the final cover shall impede the infiltration of rainwater or snow into the waste prism. The LEA says well, if no water is getting inside to make LFG then maybe you don't need to install a collection system. But gas will be generated and now with a tight cover, it can only migrate laterally or down to the groundwater.

5.0 Landfill Cover Evaluation

Once a landfill has reached the permitted volume or tonnage allowable, the landfill operator/owner is required to install a cap over the landfill to prevent surface water from infiltrating into the waste prism. There have been a few landfills in the past that incorporated a HDPE liner and soil combination but most landfill caps use a low permeable soil about 3 to 5 feet thick. The concept is to provide a barrier that will inhibit the downward migration of surface water and allow time for the water to evaporate or be drawn up by plant roots. This process is commonly known as evaportransperation.

It was believed in the 1980 & 90's that most groundwater contamination was caused by leachate and not landfill gas. However, following the installation of bottom liners the groundwater was still being impacted by landfills.

It is now being more accepted by water control boards that LFG is the primary cause of groundwater impacts. This is not surprising since LFG can migrate through most soils 1000 times faster than water.

Over the years the author has been contracted to investigate the cause of groundwater contamination from a landfill. Following PPT Power Point presentations at several water control boards it was suggested to several landfill operators that the PPT should be used to investigate groundwater impacts at their landfill.

The landfills being investigated were closed with no bottom liners and had 5 foot thick caps. They also did not have a LFG collection system installed.

The PPT was used to determine the density of the cover soil and was able to demonstrate the effectiveness of the cap by indicating the LFG being trapped under the cap. In areas where the integrity of the cap was lost, LFG pressure was indicated in the cover soil as it migrated upward. Most importantly the PPT was able to indicate the density of the native soil at the bottom of the unlined landfills. All of the landfills investigated were found to have a cap with much higher densities than the bottom native soils. It is apparent at all of these landfills that the density of the bottom native soils was not taken into account during the design of the cover soils.

Also, no LFG collection system was installed thereby not providing a pathway for the LFG to escape the landfill, it took the path of least resistance, which was through the bottom soils to the groundwater.

Unfortunately, most water control boards are still requiring caps on landfills to be designed without knowing what type of native bottom soil is present or its density/permeability. Also the water control boards are not authorized to require a collection system be installed until the groundwater is impacted. This requirement comes from the LEA or an air quality board but with a good cap and low surface emissions this requirement may not be enforced. There is a disconnect in the regulations that should be addressed. Even if the regulations do not address this problem, the engineers in the industry should automatically address it.

The condition of the bottom soils should be identified prior to designing any landfill cap and LFG vents or a collection system should be installed. The PPT is very useful in obtaining the necessary information for designing the final cap for a closed landfill.

Reminder, in flow mechanics with a constant pressure being generated it is normally found that with higher flow rates you measure lower pressures and with resistance to flow you measure higher pressures. Please keep this in mind as you read this section.

The PPT can also be used to evaluate the current conditions of a closed landfill and its in-place cover soil as demonstrated in the following figures. Figure 5-1 is from a landfill in southern California with impacted groundwater, indicates that the cover soil has a tip resistance of almost 200 tsf, which is much denser than the refuse under it. The gas pressure is indicated to increase at about 4 feet below ground surface and continues with depth. It appears that the cover soil is inhibiting the LFG flow through the cover soil due to density.

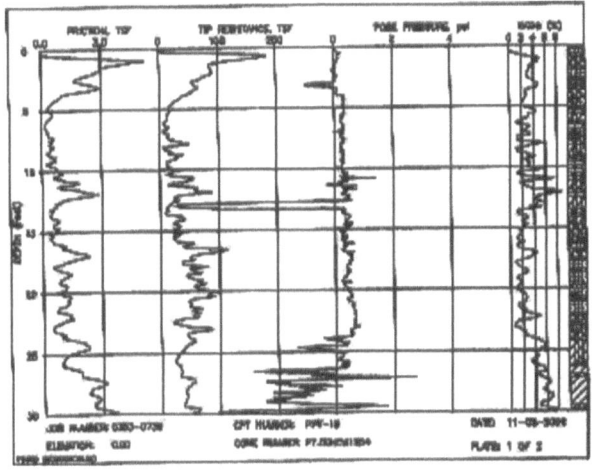

Figure 5-1

The native soil used for the cover is basically silty sand. The native bottom soil was indicated at about 27 feet bgs with a tip resistance of 50 tsf, which is much lower than the cap density. The LFG pressure being generated in the waste zone will take the path of least resistance and as the resistance drops so does the pore pressure drop, the gas is free to flow downward as indicated in Figure 5-1.

67

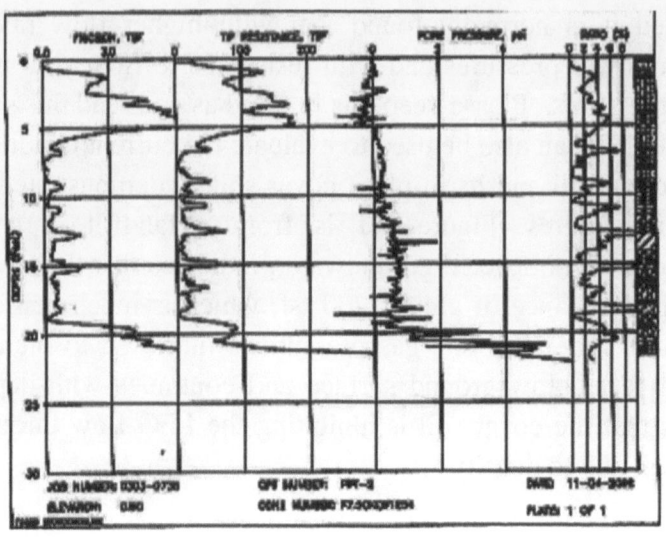

Figure 5-2

Figure 5-2 is from the same landfill but the sounding is located near a rock slope at one end of the landfill. The PPT indicates that the cover soil is also very dense at almost 200 tsf and is inhibiting the gas flow through the cap. However, the PPT also indicates that the bottom density is very high exceeding 300 tsf, as bedrock is encountered. As the PPT penetrates the bedrock the moisture in the bedrock is compressed and indicates very high pore pressure. This is PPT induced pressure and will stabilize to the upper gas pressure when the hydraulic force is released. The significant factor of this sounding is that the top and bottom of the landfill are well sealed with little apparent up or down migration of gas.

However, the gas pressure is close to that of other PPT soundings in the area. It is apparent that this gas is moving laterally to an area of the landfill where it can migrate downward. By eliminating this sounding due to the high bottom tip resistance, the other PPT soundings in this area with low, bottom tip resistance readings may be the pathway to the groundwater.

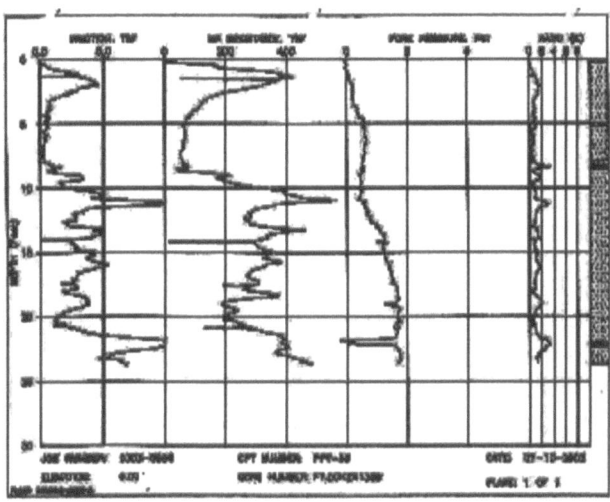

Figure 5-3

Figure 5-3 is a PPT sounding performed at another landfill in southern California, which was a trench type landfill. The cover soil was a blend of bentonite clay and native silty sand and well compacted as indicated with a tip resistance of 400 tsf. The refuse in this area was only about 5 feet thick. The bottom soil is indicated to be around 400 tsf as well but it is indicating that it is applying more resistance to the gas flow creating higher pore pressures. Although the bottom soil is very dense the LFG migrated over 300 feet down and impacted the groundwater. Several push-in vents were installed in the landfill, to relieve the gas pressure.

Chapter 6

Due to my understanding of the dynamic conditions inside landfills from performing in-situ testing in landfills the Environmental Protection Agency (EPA) asked me to review a few leachate re-circulation Bioreactor applications for the XL Program in 1999. Upon the review of these applications and their references to the short-comings of this process it became clear to me that steam injection was the way to go.

6.0 Bioreactor Landfills with Steam Injection

Current regulations do not address injecting a gas (steam) into a landfill, only liquids. Although steam condenses into liquid it should not fall under liquid injection regulations. The flow mechanics of steam and its effect on the trash prism is very different from liquid. Liquid flow is mainly controlled by gravity; steam is controlled by expansion pressure and temperature differential. Liquid will move basically downward, steam moves in all directions but basically upwards. It is suggested that steam injection should have its own set of regulations, separate from liquids.

The purpose of a Bioreactor Landfill and to re-circulate leachate and condensate is to introduce moisture, which increases the humidity in the landfill, thereby increasing the rate of biodegradation.

This will also increase the production of methane gas and to increase the settlement of the refuse by reducing the amount of organic material in the trash prism. It is also a cost effective way to dispose of leachate and condensate.

6.1 Steam Injection Landfill Bioreactors

Landfill Bioreactors have increased in popularity recently due to their ability to reduce the organic content of MSW landfills, which increases settlement (providing the re-use of the airspace) and to increase the production of methane gas for the generation of electricity. They also reduce the time the landfill may be a hazard to the environment and allow the landfill to be redeveloped for other purposes. Another concern is that bottom liners may only last about 20 years so it is important that the refuse is stabilized in a shorter period of time before the possibility of bottom liner failure.

Most Landfill Bioreactors are created by introducing liquid to the refuse, which increases biodegradation and produces methane. The main purpose of introducing moisture into a landfill should be to increase the humidity in order to accelerate biodegradation, without saturating the refuse. The liquid is currently introduced through various distribution methods, (i.e. surface flooding, sprinkling, or injection wells). However, a key problem with using liquid to attempt to raise the humidity is being able to achieve uniform distribution without saturating portions of the refuse.

Unfortunately water will not move unless the area it has made contact with is saturated to the point where the waste can't hold any more liquid.

Saturating portions of the refuse results in several concerns:

1. Increased leachate generation.
2. Regulators require additional liner protection.
3. Uneven loading on slope liners requiring over-designing.
4. Gas is bound up in liquids making extraction difficult.
5. Liquids carry suspended solids and calcium carbonate, which can block bottom drains.
6. It adds material to a landfill instead of reducing it.
7. Gas collectors become flooded.

Currently, the amount of liquid to be introduced into a Landfill Bioreactor is typically based on the field capacity (i.e. the moisture content above which the liquid will begin to drop out under gravity forces) of the refuse. This can be as high as 40% by volume and is not well known or easily defined for MSW. Unfortunately, as the organic material biodegrades the overall capacity of the refuse to hold liquid reduces, releasing the liquid and allowing it to migrate.

Even with all these challenges the Landfill Bioreactor is still more viable than the dry tomb type of landfill, for disposing of the organic threat to the environment. Every generation will always face the challenge of how to best handle their solid waste.

Every generation should strive to develop ways to eliminate the environmental hazard of their own solid waste and not leave it for the next generation to handle. Also, we should be taking advantage of the electric generating potential now when we need it the most. Therefore, a better method to raise the humidity in the refuse is needed, which will biodegrade our solid waste in our generation.

A method, which would clearly and significantly improve the "moisture distribution", while minimizing the concerns listed above, would be to change the water into steam prior to injecting it into the landfill. The first impression some people have about steam is that it will condense to liquid as soon as it leaves the injector well. However the laws of physics dictate otherwise. The steam turns to vapor then humidity and will only turn to liquid if the temperature drops below the Dew Point. Controlling the flow rates of the injection wells and the collectors and the expansion pressure of steam will extend the influence of the steam.

An example I like to use to explain this process is when someone takes a shower. The steam vapor from the hot shower does not fall on the floor as water as soon as it leaves the shower stall.

It floats around the room and will eventually condense on the walls, ceiling and floor. However, if you open the door the vapor will move down the hall and trough the house and eventually condense on walls and windows etc.

72

But if a fan is placed in a window at the far side of the house the vapor will move further through the house before condensing. The same thing happens when using gas collectors to move the steam through the waste.

Liquid flow is mainly controlled by gravity, which is downward and then laterally under a head pressure. Steam flow is controlled by expansion pressure; temperature differential, as well as wet goes to dry and moves in all directions. Therefore, the injected steam will expand throughout a much larger area within the waste and raise the humidity in the prism using much less water.

Steam expands 1,600 to 16,000 times its original volume, depending on its expansion potential (i.e. temperature), therefore requiring only a fraction of the water to achieve widespread coverage within the waste prism compared to application of water at the surface. The expansion potential is controlled by the rate of injection verses the amount of vacuum being applied by the LFG collectors. This will minimize the potential of the liquids migrating to the bottom of the landfill and to the groundwater.

6.2 Temperature Control

It is well known that temperature is a very important part of biodegradation of organic matter. Liquid injection cools the refuse and inhibits biodegradation until the temperature recovers. In the fall and winter months the liquid being injected can be very cold and will actually refrigerate the refuse. Steam will preheat the refuse and enhance biodegradation immediately. Decomposition is most active when the temperature is maintained around 120 degrees F.

Some of the refuse within a few feet of the injector well will be heated above the optimum temperature and this will inhibit biodegradation, but when the steam is reduced once the humidity has increased to the desired level, biodegradation will continue. Maintaining a higher internal landfill temperature will also convert free liquids inside the landfill into vapor and increase humidity.

Increasing the flow rate at nearby collectors will also help control temperatures near the injectors.

6.3 Enhanced Settlement, Airspace Recovery

When organic material decomposes it creates voids in the landfill. If liquid is pumped into these voids it could inhibit settlement because liquid is not compressible. Steam will flow through these voids; it will increase the rate of biodegradation/settlement and recover more airspace. Also, steam will not carry suspended solids and calcium carbonate, which can plug up bottom drains.

There is little chance of liquids exceeding the maximum head pressures allowed on geosynthetic liners. Steam injection will not create non-uniform loading of liners so when settlement does occur, less stress is placed on bottom/slope liners. The settlement is caused by removal of organic waste not by adding more weight from water.

In general, landfills are constructed using the "dry tomb method," which is keeping the landfill as dry as possible during construction and also when the landfill is closed and capped.

The reason for this is to minimize the possibility of leachate leaking to the groundwater and contaminating it.

However, the dry conditions do not allow the organic refuse to decompose and it will remain dormant for decades until water infiltrates naturally into the landfill in an uncontrolled manner, creating conditions for gas generation. Since the gas generation is uncontrolled, gas migration may result.

It was once believed that landfills do not readily give up liquids once they are introduced into the trash prism as long as the moisture content has not exceeded the field capacity of the refuse. Groundwater contamination is usually from landfill gas migration. Currently, regulatory agencies allow some leachate and condensate to be applied to the refuse as dust control while the refuse is being placed. Moisture is very important in decomposition activity.

During anaerobic decomposition, methane gas is produced which can be used as fuel to produce electricity or converted to ethanol or methanol. The more moisture in the refuse the higher the humidity, therefore more methane gas is produced. If the refuse is flooded with water the gas is bound up in the liquid and this makes it difficult to recover the methane gas.

Two factors must be considered when applying liquid to the refuse during construction. Is there a gas collection in place to collect the gas that is going to start being generated very soon and is it going to be flared or co-generated for power? Would it be more productive to leave the refuse dry until the collection system is in place and then add moisture to the refuse? A lot of gas is lost and wasted during the construction phase of the landfill, instead of producing power.

Since most closed landfills are dry tomb landfills there is probably a lot of organic material remaining to be biodegraded once moisture is introduced. Introducing water into a landfill cools the refuse. Decomposition is most active when the temperature is maintained around 120 degrees F. Introducing cool water will slow the decomposition and gas production until the temperature of the refuse recovers. The best way to increase the moisture content of the refuse by increasing the humidity is to inject live steam. This will increase the humidity in the landfill the rate of decomposition by biodegradation of the refuse, increase methane gas generation and increase the rate of settlement of the refuse in the landfill.

Again steam expands 1,600 times its original volume, therefore requiring only a fraction of water to achieve total coverage compared to application of water at the surface. For example: 1 gallon of water converted to steam will cover the same area of 1,600 gallons of water. This will minimize the potential of the liquids migrating to the bottom of the landfill and to the groundwater. Steam reduces to humidity, then to vapor and will not condense into liquid as long as the temperature remains above the Dew Point, which is the ambient air temperature outside the landfill.

This is controlled by the rate of injection verses the rate of collection from the gas collectors. The main purpose of injecting liquid into landfills is to raise the humidity of the landfill not to saturate the waste.

6.4 Steam Advantages

The following is a list of some of the advantages of using steam instead of liquids.

*   Minimize the amount of liquid being introduced into the refuse. (free liquid requires head pressures to move it through the refuse, steam is under expansion pressure)
*   Free liquid uses available airspace, steam will not.
*   Liquid is not compressible and will inhibit settlement.
*   Better moisture distribution and higher overall humidity in the refuse.
*   Steam will preheat the refuse, liquid cools the refuse.
*   Free liquids will carry suspended solids, which can plug collectors, steam will not.
*   Increase in methane production, higher BTU values and increase in flow rates.
*   Controlling the steam flow is more effective with temperature sensors than with moisture sensors.
*   Steam will not produce variable loading on slope liners, liquids can.

Another purpose of injecting steam into a landfill is to accelerate the settlement of the landfill to recover more airspace and to aid in future slope stability. In active landfills this additional airspace means more refuse can be placed on top of the landfill and delay final closure.

   As the organic material biodegrades it will create voids in the trash prism. If liquids are pumped into these voids the refuse will not settle because liquid is not compressible.

If all the liquid is not totally used up in the biodegradation of the organic material then contaminated liquid is left in the landfill with the possibility of migration to the groundwater. Steam will flow through these voids as high humidity and will allow the refuse to settle into the voids.

Currently, some bioreactor landfill permit applications are proposing that 40% by volume of liquids will be introduced into a bioreactor. It has been calculated that this amount will not exceed the field capacity of the trash prism. Unfortunately, as the organic material decomposes the field capacity reduces and the liquid begins to exceed the field capacity. Also, this means the addition of 40% more material to the landfill that will become contaminated and will take up 40% of the airspace inside the landfill, less the amount used in biodegradation. With the expansion of steam pressure, only $1/1,600^{th}$ of the 40% of the proposed liquid is required to achieve better coverage and this will never exceed the field capacity of the refuse.

Another concern of introduction of liquids into a landfill with a geosynthetic liner is slope failure due to uneven loading caused by variable head pressures. Steam injection will prevent this from happening and the heat is not high enough to harm the liner.

During a Solid Waste Association of America (SWANA) Landfill Gas Symposium in San Diego, California in 2007, several presenters that operate bioreactor landfills claimed that recent evidence has showed that only 5% of the leachate and condensate being re-circulated are retained in the waste. This means only 5% of the waste is in contact with the liquid. This indicates that liquid re-circulation is not very effective.

It appears that if the field capacity of a waste prism is 40% this means 40% of the whole landfill. However, if the liquid is applied only at the working face, which is only a small portion of the waste prism the 40% field capacity is quickly overwhelmed and flows though to the bottom drain.

77

6.5 In Situ Landfill Profiling

Prior to installing a steam injection system, it is important to know where to place the injection wells and the gas collectors. The best way to obtain this information is to perform a Piezo-Penetrometer Test (PPT) Profile on the landfill as explained in the previous chapters.

Landfills are found typically in four different weather conditions in the U. S. Hot and dry, hot and wet, cold and dry and cold and wet. The climate the landfill is in will be part of the landfill profile. Where is the best place in the country to use the steam injection process? The answer is everywhere. The main purpose of the steam is for even distribution of moisture and preheating the waste. No matter the location of the landfill all active landfills in the U.S. must be operated basically as a dry tomb landfill.

The following are the characterizations of the conditions found in region of the country:

- Hot & Dry – Waste is dry and warm with slow biodegradation and settlement. Lack of rain also means lack of water for conventional water injection bioreactor. With 1,600 times less water required with full coverage, makes steam injection more practical.
- Hot & Wet – Waste is moist and warm but landfill is graded to keep water away from working face. Waste is placed in prepared wet decks during rain events so vehicles don't get stuck. On dry days waste is placed in dry areas so the waste is the same as hot/dry landfills. Perched zones will be found in the wet areas. This liquid can be converted to steam and injected into the dry areas. (see Down Hole Boilers)
- Cold & Dry – Waste is dry and may be frozen when placed in the landfill. Temperatures inside the landfill must recover before biodegradation can begin. Steam will maintain a more consistent temperature inside the landfill. There may be a lack

of water, lower water requirements for steam make it more feasible.
- Cold & Wet – Waste is moist and cold but landfill is graded to keep water away from working face. Waste is placed in prepared wet decks during rain events so vehicles don't get stuck. On dry days waste is placed in dry areas so the waste is the same as dry landfills. Temperatures inside the landfill must recover before biodegradation can begin. No lack of water, steam is a good way to dispose of excess water. Steam will also maintain a more consistent temperature inside the landfill.

Of course if the PPT Profile indicates many perched water zones in the waste prism the application of steam will be different than a profile that indicates dryer waste conditions.

It would not be practical to install injectors and collectors into a perched layer of water, which would make the process ineffective. However, since there is water inside the landfill all that is needed is to distribute it to dryer portions of the landfill.

6.6 Down Hole Boiler

Many landfills have perched liquid layers even if they are not re-circulating leachate. Also many landfills have flooded gas collectors making gas extraction difficult. Usually these collectors have to be pumped out and the leachate treated or disposed of off-site. Although the landfill may be flooded in some areas usually in the prepared rain day wet deck areas, most of the remaining areas of the landfill will be dry.

Instead of pumping the leachate with all of the hazardous particular matter, it is more practical to install a down hole boiler down the flooded collectors and boil only the water out of the collector. All of the particulate matter stays in the landfill and only the steam is piped to a dryer portion of the landfill.

The injector for the steam is placed in close proximity to a gas collector in the dryer portion of the landfill. The vacuum influence of this collector is used to pull the steam into the dry refuse.

The flooded collector would be connected to the vacuum header with a vacuum applied even though at this time the screens are probably under water. By lowering the atmospheric pressure on the water it will boil much more aggressively. Also by adding a handful of salt down the well will boil the water faster.

A separate smaller pipe that is connected to the boiler chamber transmits the steam up, out of the flooded collector and to the steam injector under the vacuum influence of the other dry collector. Once the water level is lower and the screens are exposed gas will begin to enter the collector. The vacuum will have to be increased to keep a negative pressure on the flooded collector. The water level must always be above the boiler chamber or gas will be pulled back through the steam line. A check valve in the steam line can prevent this from happening. Also a float switch above the boiler will turn off the boiler when the water level gets too low.

You may be wondering what keeps the boiler from melting a PVC or HDPE collector.

As figure 6-1 illustrates the boiler chamber is surrounded by a shroud to provide another layer of water for cooling the shroud and protecting the collector casing. If the float switch fails to shut off the boiler and it runs dry it could melt the casing. Float switches have been found to be reliable but nothing is infallible. The system should be checked periodically as you would a pump.

Currently the regulations do not address this process at all so if permitting is required, it is not clear. However as to not evoke the ire of local agencies it would be best to discuss this with them first.

Just remind them that landfill gas is saturated with water vapor and moving LFG from one area of the landfill to another is allowable under current regulations. This happens every time a flare or co-gen. is turned on or off. This steam is a vapor with very, very low methane content.

Please keep in mind that this process is also part of our patent process so please give us a call before doing this.

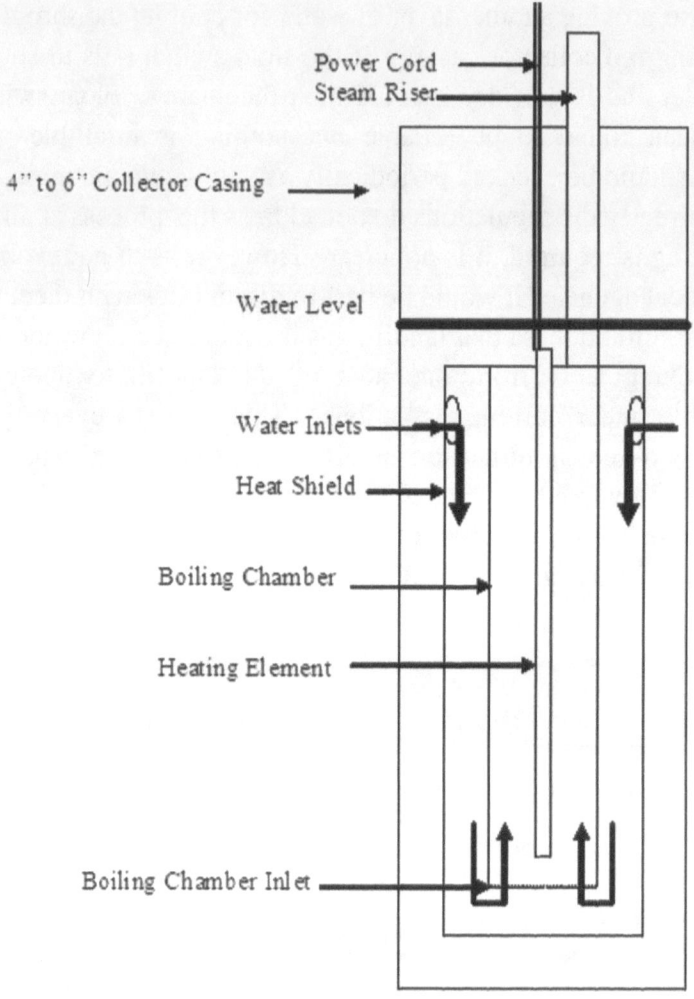

Power Cord
Steam Riser

4" to 6" Collector Casing

Water Level

Water Inlets

Heat Shield

Boiling Chamber

Heating Element

Boiling Chamber Inlet

Figure 6-1

6.7 Anaerobic Steam Injection

The following will describe the two basic types of injection methods, the anaerobic (no oxygen) steam injection method, which produces methane and carbon dioxide and aerobic steam injection method (with oxygen), produces just carbon dioxide.

Most landfills install gas collectors basically in a grid pattern about 200 feet apart from each other. So basically there is a collector at the four-corners of a box the size of an acre. If four collectors are not installed to complete the box a PPT sounding is performed to verify the location for a collector.

Following the PPT, the cone truck would install a 2" diameter steel push-in collector. At the center of the box a 2" diameter steel push-in injector would be installed. A ½" diameter steam hose would connect the injectors to the boiler or heat exchanger. The hose is buried in the cover soil to protect the hose and insulate it. A 5-foot long lateral pigtail connects the injector to the main steam line. As the landfill settles and the injector rises above the landfill surface the pigtail raises also, leaving the main steam hose buried. The injector pipes are installed in 5-foot long section of blank pipe. When 5 feet of the pipe is exposed above the landfill it is disconnected from the injector and the pigtail is reconnected to the top of the injector pipe and the process is repeated.

The PPT Profile will determine the actual depth of the screen interval, which will be above a dense layer.

In most cases where there are already existing collectors the collectors will be deeper than the injectors, which rarely go deeper than mid-depth of the landfill. The waste is too dense at deeper depths to be practical. It is possible that the drilled in plastic wells may buckle when the landfill begins to subside so the collectors will have to be replaced with steel push-in collectors which can stand up to subsidence. This is why push-in collectors should be used whenever possible.

Due to the ability of the PPT to identify dense layers, which act as a barrier layer, it is believed that steam injection can be used in old landfills that do not have a bottom liner.

With precise control of the steam by temperature sensors and follow-up PPT investigations the threat to the groundwater is minimal. The benefits of biodegrading the organic material in a few years instead of decades out ways any risk to the groundwater. With steam injection there will not be any free liquid remaining in the landfill to threaten the groundwater.

The steam can be produced in a gas-fired boiler, heat exchanger on the gas flare, or from the exhaust steam from a power plant. The steam is not produced under high pressure and high flow rates. Actually high temperature water is transmitted to the injectors and that is where the water is flashed to steam inside the landfill.

This prevents the temperatures from increasing to the point of killing the anaerobic bacteria.

Only the immediate area (~ 1 to 3 feet) around the injection wells would be affected by high temperatures. When the steam is turned off the bacteria quickly recovers. Flow of the steam through the waste prism is controlled by the amount of vacuum applied to the collectors. The steam is pulled through the waste not pushed.

If the waste is very fresh our patent also allows us to inject ammonia or urea using steam as a carrier gas. This will displace any air that may still be in the waste and neutralize the ph of the refuse. Accelerating methane generation in the waste prism.

6.8 Aerobic Steam Injection

In addition to injecting steam into an anaerobic bioreactor, steam can also be injected into an aerobic bio-cell. Aerobic digestion is more robust than anaerobic digestion; however the biggest danger to operating an aerobic bio-cell is spontaneous combustion due to air injection. To prevent combustion it is necessary to maintain high moisture content in the trash prism. Two areas most prone to combustion due to evaporation of the moisture in the trash is the point of entry of the air stream into the trash prism and the point of exit of the air stream with the removal of moisture.

The air stream migrates through voids in the trash prism that liquid can not travel due to gravity influence. The best way to replenish moisture to the trash prism is to moisturize the air stream with steam prior to injection. The air stream would then be moisturizing the refuse with more uniformity, instead of drying it out.

The warm steam will also maintain the temperature of the refuse especially during the winter months when cold air inhibits decomposition.

Other aerobic bio-cells have tried to use surface flooding to moisturize the waste but again air cannot travel through saturated waste so the water goes one way and the air takes the path of least resistance.

Some people have tried to sprits the air stream with water but the water particles are too big so they are filtered out of the air stream as it passes through the waste prism near the injection points. What occurs is a big gooey mess around the injection points and airflow is restricted.

6.9 Monitoring of the Methods

To assist in obtaining authorization from the various regulatory agencies a comprehensive monitoring system should be in place. This will also be used to make adjustments to the injection system. The PPT Profile will identify the best locations to place moisture sensors and temperature sensors. The temperature sensors will be placed to monitor the movement of the steam through the trash prism. These sensors will be more responsive than the moisture sensors and will be able to track the migration of the steam.

When the temperature of the refuse reaches a predetermined level the steam injection will be reduced. The moisture sensors will monitor the amount of liquid accumulating on the dense layer below the injection wells as a back-up system. If a liquid head is detected the amount of steam will be reduced. The PPT rig can install the sensors at very accurate depths based on the baseline PPT Profile.

Follow-up PPT Profiles will monitor the decomposition of the organic material and the settlement between the dense layers. As the volume of the organic material between the dense layers reduce, the amount of steam will also be reduced. This will help prevent any liquid accumulating on the dense layers.

The top deck elevation of the landfill is surveyed during the PPT Profiling. This data will be used to monitor the overall settlement of the landfill.

Monitoring of the landfill gas at each collector will indicate the effectiveness of the above Methods (i.e. increase in flow rates, higher CH_4 concentrations and higher BTU values). Adjustments to the Methods would be made according to these measurements.

Chapter 7

When I finally received my patent in 2002, I went on a campaign to find a landfill that wanted to pay for a Pilot Study for the Steam Injection Process. Of course the first thing I heard from Landfill operators was "where has this been done before". I would say nowhere that's why it's called a pilot study. They would say "see ya", and hang up. Or if they didn't hang up they would say you will never get the water board to allow it. Well, after a couple of years of striking out on that and dealing with shady investment types, a friend suggested that I create a LLC company and sell shares in it to raise capital for the pilot study. So I created California Steam LLC and sold 20% of the Company raising $500,000.00 for the Pilot Study. I am very grateful to these investors for believing in the process and me. Some of the investors were people who wanted to be a part of the landfill issue and the environment, some were interested in the potential return on investment when the project went full scale and some were friends and colleges in the industry that know my work and understand the process.

I heard from another friend that the Miramar Landfill in San Diego, California was considering making a Bioreactor by injecting waste water from a nearby treatment plant. It appears that the landfill is running out of airspace 11 years earlier than what was planned. As it turns out the landfill only has a 24" diameter bottom drain and it requires a 48" diameter bottom drain to be allowed consideration for a water injected Bioreactor.

So I went to the City of San Diego and presented a Power Point presentation and explained what we plan to do and that it wouldn't cost them a dime. They said sure if you can get the water board and LEA to permit it.

Then the fun began, but within two months we had all the permits required by all the agencies. Actually the agencies were very supportive of this technology. They understand that this process will eliminate an environmental hazard in years instead of decades.

7.0 Steam Injection Pilot Study

In 2005 STI performed a Steam Injection Pilot Study at the Miramar Landfill in San Diego California. Miramar Landfill encompasses 500 acres. STI performed a PPT profile on about 5 acres of the site and selected 2 acres for the study. A test cell was created in one acre, and a control cell was located adjacent to the test cell. The main objective of this pilot study was to demonstrate that steam injection would accelerate the settlement of the refuse and recover airspace. Although the methane quality increased the total volume of the gas generated was not monitored. It was more important that the vacuum flow was controlled to maximize the migration of the steam through the waste prism and not the maximum flow of LFG out of the collectors. The following is the rest of the story.

7.1 Introduction

In August, 2004 STI Engineering presented a proposal to the City of San Diego, Refuse Disposal Division to perform a Steam Injection Pilot Study at the Miramar Landfill, located in San Diego, California. On May 5, 2005 STI Engineering received the Right of Entry from the City of San Diego to perform the Steam Injection Pilot Study. On May 16, 2005 STI Engineering mobilized a PPT rig and a Geoprobe unit to the subject site.

The 30 ton PPT rig that was available at the time of the start date did not have a large enough center hole to accommodate the 2" diameter pipe used for the injectors and collectors, so the Geoprobe was used to install the pipes.
No drilling was used to install any of the instruments in the test area minimizing the disturbance to the test area.

The following outlines the project scope of work, describes the field operations and provides interpretations of the PPT data.

7.2 Scope of Work

The scope of work completed consisted of the following tasks:

- Developed a Health and Safety Plan for the proposed work on the subject site.
- Pre-qualified selected locations on the landfill using the PPT prior to installing 2-inch diameter steel push-in collectors. The PPT data was used to indicate the following:
  - Verify the presence and density of refuse.
  - Determine if LFG pressure was present.
  - Determine the best depths to install the screen sections of the injectors and extraction wells.
  - Verify that no liquid layers were present that could impact the injectors or extraction wells.
- Installed 6 schedule 80 black steel 2" diameter extraction wells (EW) with (1/8-inch mill slot screen) for gas in the test area.
- Installed 2 schedule 80 black steel 2" diameter extraction wells (EW) for gas in the control area.
- All extraction wells were connected to a 2" diameter Landtec well heads and then to the vacuum header.
- Installed 3 steam injectors using 2" diameter steel schedule 80 black pipe.
- Installed 9 thermocouples into the refuse at various depths and at various locations in the test area. Installed 1 thermocouple on each of the 3 steam injectors.
- Installed 2 Static Piezometers in the test area.
- Install 5 settlement monuments in the test area and one in the control area.
- Prepared and submitted this progress report.

A Health and Safety Plan was developed prior to the start of the field operations and submitted to the City of San Diego. A Health & Safety meeting was conducted with all parties involved in the field activities on May 15, 2005 prior to the start of the work.

89

A Rae Q-Rae Lower Explosive Limit (LEL) meter was used to monitor the air inside the PPT rig. The readings never went beyond the background levels at the landfill site. Operations were conducted in Level D protection.

7.3 Field Operations

The field operations began by performing several PPT soundings on a 200-foot grid to determine the overall conditions of the landfill waste prism. The grids were reduced to 100 feet and then to 50-feet to obtain enough data in the selected test area for a closer look at the conditions in the study area.

Once a 200' x 200' area was selected based on the PPT data, 6 push-in extraction wells were installed, 3 along the eastside of the test area and 3 along the west side of the test area. (see Plan View Figure 7-1)

About 200 feet east of the test area, 2 push-in extraction wells were installed in a control area to monitor the gas production by natural processes.

Three push-in steam injectors were installed along the centerline of the test area. The screen sections (1/8-inch mill slot) varied from 40 feet to 20 feet below ground surface. (see Plan View Figure 7-1)

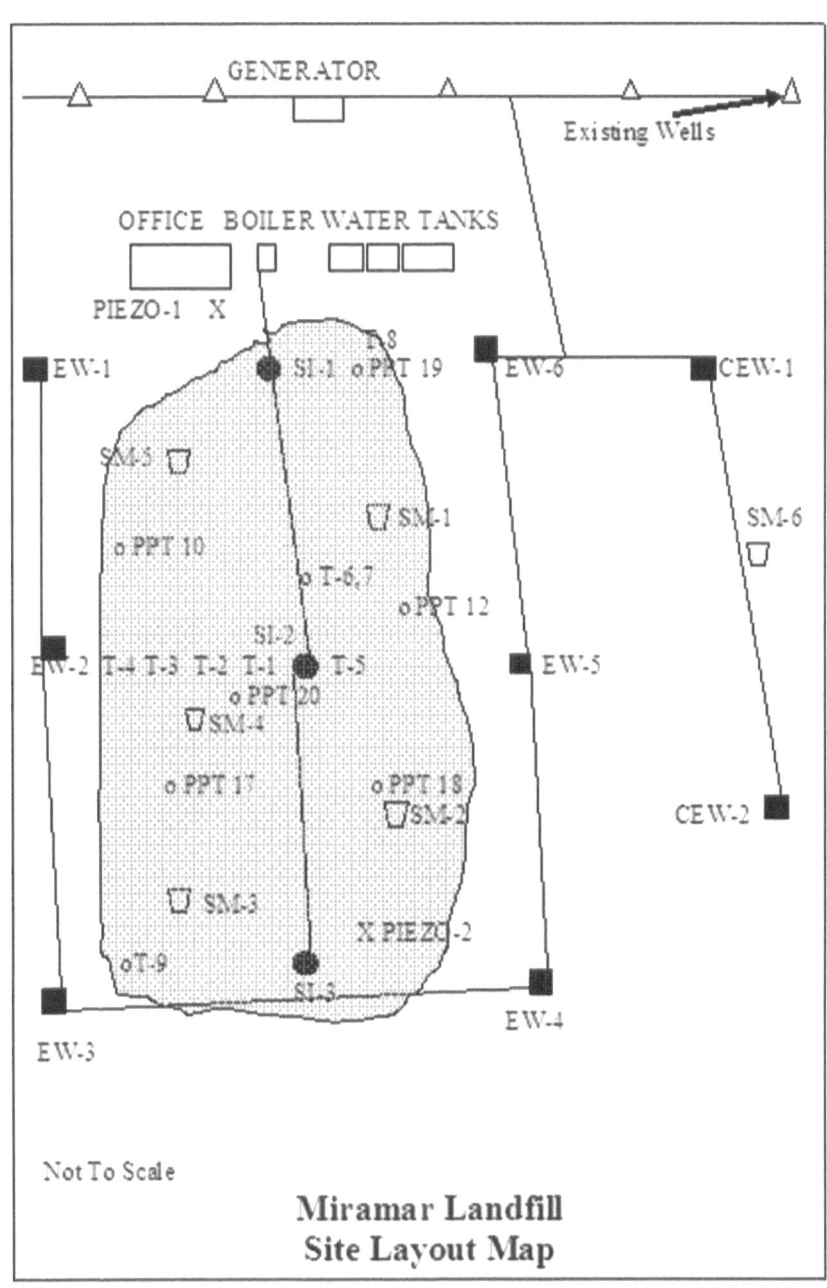

Plan View
Figure 7-1

91

Any PPT hole created by a sounding that was not used for a pilot hole for a collector/injector was sealed with Bentonite following each PPT sounding. The base of each collector, injector and Piezometers at the ground surface was sealed with Bentonite.

Two Static Piezometers were installed, one at the north side of the test area and one at the south side of the test area. The five foot slotted section was installed at a depth of 47 to 52 feet below ground surface. (see Plan View Figure 7-1)

Using the PPT rig, 9 thermocouples were installed in the test area. Four of the thermocouples were installed at a depth of 22 feet bgs and at 25-foot intervals between SI-2 and EW-2. (see Plan View Figure 7-1) These thermocouples will be used primarily to monitor the migration of the steam from the injector to the collector.

The other thermocouples were installed at various locations throughout the test area to monitor the coverage of steam. A thermocouple was attached to each of the 3 steam injectors. All of the thermocouples were connected to an Omega scanning data logger located inside a data acquisition trailer on site.

Five settlement monuments consisting of an 18" long rebar driven into the cover soil and secured in place by concrete were installed in the test area (see Plan View Figure). One settlement monument was installed in the control area. The baseline survey of the settlement monuments, the PPT locations, the steam injector locations, the thermocouple locations and the Piezometer locations was performed on May 25, 2005.

7.4 Data Interpretation

The following provides interpretation of the PPT soundings performed at the Miramar Landfill. Table 1 summarizes the data. The PPT locations were designated as EW for extraction well if a collector was installed following the PPT sounding.

The PPT locations were designated as SI for steam injector if a steam injector was installed at that location. Also the PPT locations were designated as Piezo if a Piezometer was installed at that location. A key function of the PPT is to interpret soil type through established methods of relating tip and sleeve resistance. Refuse is also distinguished by the instrument readings.

Pore pressure measurements are recorded in real time as the instrument was hydraulically pushed through the landfill. Various procedures were performed to differentiate between liquid and gas pressures. Some perched liquid was indicated in PPT-4.

PPT soundings were performed at 50', 40' and 25' from the existing collectors adjacent to the test area. Significant vacuum was only indicated in the sounding that was only 25 feet from the collector.

Based on this information the test area was placed at least 100 feet away from the line of existing collectors to minimize their influence on the test area.

The following paragraphs summarize the soundings and installations at each PPT location in the sequence of completion.

PPT-1

The location of this sounding was 200 feet south of well 27A. The cover soil is 5 feet thick. Low gas pressure was indicated at 5 to 37 feet below ground surface (bgs) from 37 to 79 feet bgs gas pressure increased to 10 psi. No liquid was indicated in this sounding. The sounding was terminated at 79 feet bgs.

PPT-2

The location of this sounding was 100 feet south of well 27B. The cover soil is 7 feet thick. Gas pressure was indicated at 7 to 37 feet bgs. No liquid was indicated in this sounding. The sounding was terminated at 37 feet bgs.

PPT-3

The location of this sounding was 200 feet south of well 27C. The cover soil is 7 feet thick. Gas pressure was indicated at 7 to 33 feet bgs. No liquid was indicated in this sounding. The sounding was terminated at 33 feet bgs.

PPT-4

The location of this sounding was 200 feet south of well 27D. The cover soil is 6 feet thick. Gas pressure was indicated at 6 to 18 feet bgs. From 18 to 22 feet bgs no gas pressure was indicated. From 22 to 60 feet bgs gas was indicated. Liquid was indicated at 58 to 60 feet bgs. The sounding was terminated at 62 feet bgs.

PPT-5

The location of this sounding was 100 feet south of well 27E. The cover soil is 8 feet thick. Gas pressure was indicated at 8 to 20 feet bgs. From 20 to 52 feet bgs it appears that the gas is under the influence of the nearby collector. From 52 to 60 feet bgs gas was indicated. No liquid was indicated in this sounding. The sounding was terminated at 60 feet bgs.

PPT-6

The location of this sounding was 50 feet south of well 27D. The cover soil is 7 feet thick. Gas pressure was indicated at 7 to 23 feet bgs. No liquid was indicated in this sounding. The sounding was terminated at 23 feet bgs.

PPT-6A

The location of this sounding was 40 feet south of well 27D. The cover soil is 7 feet thick. Gas pressure was indicated at 7 to 23 feet bgs. No liquid was indicated in this sounding. The sounding was terminated at 23 feet bgs.

PPT-7

The location of this sounding was 25 feet south of well 27C. The cover soil is 6 feet thick. Gas pressure was indicated from 6 to 18 feet bgs. Vacuum was indicated from 18 to 23 feet bgs and at 35 to 51 feet bgs. Gas pressure was also indicated at 32 to 35 feet bgs. No liquid was indicated in this sounding. The sounding was terminated at 70 feet bgs.

PPT-8 & EW-1

The location of this sounding was 100 feet south of well 27D. The cover soil is 7 feet thick. Gas pressure was indicated from 7 to 30 feet bgs and from 32 to 68 feet bgs. No gas pressure was indicated from 30 to 32 feet bgs. No liquid was indicated in this sounding. The sounding was terminated at 68 feet bgs. A push-in extraction well was installed at this location.

PPT-9

The location of this sounding was 150 feet south of well 27D. The cover soil is 7 feet thick. Gas pressure was indicated from 7 to 52 feet bgs. No liquid was indicated in this sounding. The sounding was terminated at 52 feet bgs.

PPT-10

The location of this sounding was 150 feet south of well 27D and 50 feet east of PPT-9. The cover soil is 7 feet thick. Gas pressure was indicated from 7 to 53 feet bgs. No liquid was indicated in this sounding. The sounding was terminated at 53 feet bgs.

PPT-11

The location of this sounding was 150 feet south of well 27D and 50 feet east of PPT-10. The cover soil is 7 feet thick.
Gas pressure was indicated from 7 to 57 feet bgs. No liquid was indicated in this sounding. The sounding was terminated at 57 feet bgs.

PPT-12

The location of this sounding was 200 feet south of well 27C and 50 west. The cover soil is 8 feet thick. Gas pressure was indicated from 8 to 54 feet bgs. No liquid was indicated in this sounding. The sounding was terminated at 54 feet bgs.

PPT-13 & EW-5

The location of this sounding was 200 feet south of well 27C. The cover soil is 7 feet thick. Gas pressure was indicated from 7 to 20 and from 35 to 50 feet bgs. Vacuum was indicated 20 to 35 feet bgs. No liquid was indicated in this sounding. The sounding was terminated at 50 feet bgs.

PPT-14

The location of this sounding was 300 feet south of well 27C. The cover soil is 7 feet thick. Gas pressure was indicated from 7 to 56 feet bgs. No liquid was indicated in this sounding. The sounding was terminated at 56 feet bgs.

PPT-15

The location of this sounding was 300 feet south of well 27C and 100 feet west of PPT-14. The cover soil is 6 feet thick. Low gas pressure was indicated from 6 to 23 feet bgs. No liquid was indicated in this sounding. The sounding was terminated due to refusal at 23 feet bgs.

PPT-16

The location of this sounding was 300 feet south of well 27D. The cover soil is 6 feet thick. Low gas pressure was indicated from 7 to 53 feet bgs. No liquid was indicated in this sounding. The sounding was terminated due to refusal at 53 feet bgs.

PPT-17

The location of this sounding was 200 feet south of well 27D and 50 feet east of EW-2. The cover soil is 8 feet thick. Gas pressure was indicated from 8 to 16 and at 32 to 56 feet bgs. No liquid was indicated in this sounding. The sounding was terminated at 56 feet bgs.

PPT-18

The location of this sounding was 200 feet south of well 27C and 50 feet west of EW-5. The cover soil is 8 feet thick. Gas pressure was indicated from 8 to 50 feet bgs. No liquid was indicated in this sounding. The sounding was terminated at 50 feet bgs.

PPT-19

The location of this sounding was 100 feet south of well 27C and 50 feet west of EW-6. The cover soil is 8 feet thick. Gas pressure was indicated from 8 to 49 feet bgs. No liquid was indicated in this sounding. The sounding was terminated at 49 feet bgs.

Gas Extraction Wells

EW-1 & PPT-8

The location of this sounding was 100 feet south of well 27D. The cover soil is 7 feet thick. Gas pressure was indicated from 7 to 30 feet bgs and from 32 to 68 feet bgs. No gas pressure was indicated from 30 to 32 feet bgs. No liquid was indicated in this sounding. The sounding was terminated at 68 feet bgs. A push-in extraction well was installed at this location.

EW-2

The location of this sounding was 175 feet south of well 27D. The cover soil is 5 feet thick. No gas pressure was indicated in this sounding. No liquid was indicated in this sounding.

The sounding was terminated at 53 feet bgs. A push-in extraction well was installed at this location.

EW-3

The location of this sounding was 225 feet south of well 27D and 50 feet east. The cover soil is 5 feet thick. Gas pressure was indicated at 5 to 44 feet bgs. No liquid was indicated in this sounding. The sounding was terminated at 44 feet bgs. A push-in extraction well was installed at this location.

EW-4

The location of this sounding was 225 feet south of well 27C and 50 feet west. The cover soil is 5 feet thick. Gas pressure was indicated at 5 to 74 feet bgs. No liquid was indicated in this sounding. The sounding was terminated at 74 feet bgs. A push-in extraction well was installed at this location.

EW-5 & PPT-13

The location of this sounding was 225 feet south of well 27C and 50 feet west. The cover soil is 5 feet thick. Gas pressure was indicated at 5 to 74 feet bgs. No liquid was indicated in this sounding. The sounding was terminated at 74 feet bgs. A push-in extraction well was installed at this location.

EW-6

The location of this sounding was 100 feet south of well 27C and 50 feet west. The cover soil is 7 feet thick. Gas pressure was indicated at 7 to 60 feet bgs. No liquid was indicated in this sounding. The sounding was terminated at 60 feet bgs. A push-in extraction well was installed at this location.

Control Cell Extraction Wells

Two push-in steel collectors were installed 100 feet east of the test site to monitor the natural processes occurring in the landfill. This baseline information will be used to compare the differences in gas production and settlement in the test area.

CEW-1

The location of this sounding was 100 feet south of well 27B and 100 feet west. The cover soil is 8 feet thick. Gas pressure was indicated at 8 to 50 feet bgs. No liquid was indicated in this sounding. The sounding was terminated at 50 feet bgs. A push-in extraction well was installed at this location.

CEW-2

The location of this sounding was 300 feet south of well 27B and 100 feet west. The cover soil is 6 feet thick. Gas pressure was indicated at 6 to 54 feet bgs. No liquid was indicated in this sounding. The sounding was terminated at 60 feet bgs. A push-in extraction well was installed at this location.

Steam Injection Wells

SI-1

The location of this sounding was 100 feet south of well 27C and 100 feet west. The cover soil is 8 feet thick. Gas pressure was indicated at 8 to 18 and 30 to 50 feet bgs. Some Vacuum influence was indicated at 18 to 30 feet. No liquid was indicated in this sounding. The sounding was terminated at 50 feet bgs. A push-in steam injector was installed at this location.

SI-2

The location of this sounding was 175 feet south of well 27C and 100 feet west. The cover soil is 7 feet thick. Gas pressure was indicated at 7 to 56 feet bgs. No liquid was indicated in this sounding. The sounding was terminated at 56 feet bgs. A push-in steam injector was installed at this location.

SI-3

The location of this sounding was 250 feet south of well 27C and 100 feet west. The cover soil is 6 feet thick. Gas pressure was indicated at 6 to 51 feet bgs. No liquid was indicated in this sounding. The sounding was terminated at 51 feet bgs. A push-in steam injector was installed at this location.

Piezometers

Two static Piezometers were installed at the north end and the south end of the test site. The 5 foot slotted section was in stalled just above a dense layer indicated in the PPT soundings. The purpose of the Piezometers is to monitor the moisture conditions of the test area. If liquid is detected in the Piezometers the amount of steam being injected will be adjusted. Currently, as of June 6, 2005 no liquid is detectable by a well sounder in either Piezometer.

Piezo-1

The location of this sounding was 100 feet south of well 27C and 125 feet west. The cover soil is 7 feet thick. Gas pressure was indicated at 15 to 60 feet bgs. No liquid was indicated in this sounding. The sounding was terminated at 60 feet bgs. A 1" PVC Piezometer with a 5 foot slotted section was installed at this location.

Piezo-2

The location of this sounding was 275 feet south of well 27C and 75 feet west. The cover soil is 8 feet thick. Gas pressure was indicated at 24 to 60 feet bgs. No liquid was indicated in this sounding. The sounding was terminated at 60 feet bgs. A 1" PVC Piezometer with a 5 foot slotted section was installed at this location.

7.5 Conclusions & Goals

- The PPT data indicated that the refuse in the top 40 feet of the landfill has a density of 25 to 50 tons per square foot (tsf) and that the soil layers have a density of about 100 tsf with peaks of tip resistance of 300 tsf. At depths of about 45 feet bgs the density of the refuse increases to about 100 tsf to 150 tsf with depth. It will be our goal to match this density with the density of the top 40 feet of the landfill with the steam injection process if the waste is allowed to settle. If bridging occurs then the density will lessen as the organic material biodegrades.
- Some vacuum influence was indicated in some of the soundings within 100 feet of the existing collectors. It appears that the amount of vacuum available in this area is not keeping up with the gas generation rate. The PPT sounding located 25 feet from 27C did indicate negative pressures. The first screening soundings indicated no vacuum in the test area, but as more soundings were performed and as collectors/injectors were installed, it appears the vacuum influence was being extended into the test area by the holes punched by the PPT cone.
- Only PPT-4 indicated a thin perched liquid layer. It appears the thick cover soil layer and good surface grading has minimized the amount of rainwater infiltrating the waste prism.
- Currently, the thermocouples are indicating that the landfill is at around 96° F. It will be our goal to increase the refuse temperature to 120° F, for maximum degradation.

TABLE 1

Summary of PPT Soundings and EW Installations

PPT Number	Percentage Of Refuse	Percentage Of Soil	Depth Of Screen	Total PPT Depth
	(50 ft column)	(50 ft column)	(feet bgs)	(feet bgs)
PPT-1	84	16	0	79
PPT-2	84	16	0	32
PPT-3	63	37	0	32
PPT-4	70	30	0	62
PPT-5	70	30	0	60
PPT-6	65	35	0	23
PPT-6A	70	30	0	70
PPT-7	80	20	0	70
PPT-8	70	30	0	68
PPT-9	70	30	0	52
PPT-10	60	40	0	54
PPT-11	50	50	0	53
PPT-12	60	40	0	54
PPT-13	70	30	0	50
PPT-14	70	30	0	56
PPT-15	65	35	0	23
PPT-16	70	30	0	53
PPT-17	70	30	0	52
PPT-18	60	40	0	50
PPT-19	70	30	0	48
Collectors				
EW-1 PPT-8	70	30	20 to 35	Repeat
EW-2	60	40	25 to 40	53
EW-3	55	45	15 to 30	45
EW-4	70	30	20 to 35	74
EW-5/PPT-13	70	30	20 to 35	Repeat
EW-6	70	30	20 to 35	60

TABLE 1

Summary of PPT Soundings and EW Installations
(continued)

PPT Number	Percentage of Refuse	Percentage of Soil	Depth of Screen	Total PPT Depth
	(50 ft column)	(50 ft column)	(feet bgs)	(feet bgs)
CEW-1	50	50	20 to 35	50
CEW-2	70	30	17 to 37	60
Steam Injector				
SI-1	70	30	30 to 45	55
SI-2	60	40	20 to 35	56
SI-3	70	30	20 to 35	50
Piezometers				
Piezo-1	60	40	47 to 52	52
Piezo-2	70	30	47 to 52	60
Thermocouples				
T-1				22
T-2				22
T-3				22
T-4				22
T-5				31
T-6				15
T-7				35
T-8				15
T-9				31
T-10 – SI-1				Top of Inj.
T-11 – SI-2				Top of Inj.
T-12 – SI-3				Top of Inj.
Total Footage			**180**	**1,871**

Chapter 8

After a few months of testing we submitted this progress report to the City and the agencies.

8.0 Introduction

This report is the second Interim Report for the Steam Injection Pilot Study being performed at the Miramar Landfill in San Diego, California. Field investigations began on May 16, 2005 with the PPT subsurface profiling and installation of the LFG collectors, injectors and monitoring instrumentation. On June 15, 2005 the steam boiler was brought on line and the steam process began by injecting 700 gallons of landfill leachate water the first day. An Interim Report (Reference) was presented to the City, describing the investigation process and results, as well as outlining the locations and depths of the collector wells and injector wells. Since that time the Pilot Study has continued, as described below:

8.1 Progress

In order to optimize conditions for landfill volume reduction by bio-degradation, it is imperative to provide sufficient steam to the test prism such that virtually all voids can be filled with moisture. STI steam injection operations began at the site using a 360,000 BTU boiler capable of delivering over 4,000 gallons of water converted to steam per 24-hour period.

The boiler performance was considered as optimum based on calculations regarding the amount of void space within the total waste volume in the test area, approximately 20%.

The water supplied for the steam injection comes from the landfill leachate collection system and as a result, has a very high amount of both particulate and dissolved solid matter suspended within. This water introduced these solids to the boiler and injection system causing significant mechanical problems, machinery clogs, and resulting down time to clean and repair equipment. The degree of filtration was not apparent or planned for in the initial design of the system.

Filtering the water is the only viable solution, but has been hampered by the extremely high concentrations of solid particles in the water over-loading all devised filtration systems. As a result, one of the least desirable conditions began to occur with greater frequency, down time. The system could not operate for the planned extended periods due to the maintenance of filtration and injection equipment. However, the majority of the filtration problems have been overcome and the system is functioning more reliably.

During the last 4 months only about 150,000 gallons of water has been injected as steam, which is well below the 450,000 gallons initially planned. At this time the landfill is only producing about 1,500 gallons of leachate and condensate per day, insufficient to achieve the planned objectives, whereas, the steam injection process requires 3,300 gallons per day, yielding a water shortage of 1,800 gallons per day.

A request has been submitted to the City to use locally available recycled water as a supplement to the leachate water, allowing for production goals to be achieved.

During the past 4 to 5 months, temperature readings have been recorded, piezometer readings taken, and LFG readings were measured using a Landtec Gem 2000. The chronological results of the various measurements have been tabulated are included at the end of this report.

Generally, all aspects show favorable increases during the monitoring period, despite the delays and inconsistent operation times.

8.2 Current Operations

The problems encountered due to the high concentrations of solid particles within the landfill leachate water being used in the injection system have resulted in significant cost increases and delays.

The Pilot Study is now about 2-months behind planned schedule and was expected to be nearly concluded as of this date. The impact of this delay on the Study will essentially be to require continuation of operations into the upcoming rainy season. As a result, earthen berms are being constructed around the test site to control surface water drainage.

In an effort to overcome the past delays it will be necessary to increase steam injection into the refuse prism. There are two primary tasks required to achieve this condition; increase the available water for steam injection, and, increase the number of steam injection points.

A request has been made through the City and other governmental agencies involved, allowing for the use of recycled water as a supplement to the landfill leachate water. The sooner this approval is granted, the sooner full production rates can be achieved. Another large water storage tank will be acquired once approval for the recycled water is granted, in order to keep a plentiful reserve of water available for steam injection as necessary. This will improve system performance by maintaining longer-term operation times.

In addition, the extraction wells were constructed with steel piping to allow them to be converted into injection wells if necessary.

Two of the steam boilers have been connected to EW-2 and EW-5, making those two wells potential steam injection points. The steam hoses have been buried under the cover soil for protection and insulation. When enough water is available and the system is fully operational, an evaluation will be made regarding the benefit of bringing these two wells into the injection system to increase the steam migration throughout the test site.

STI is simulating the current LFG collection system found at the site by operating only the 4 corner collectors at the test cell. In a full-scale scenario STI would set a steam injection well in the middle of each acre with a collector at each corner. The corner collectors would pull the steam across the treated acre. The current test scenario is proving to be effective at this time.

However, we are injecting steam from 3 injectors instead of one to make up for lost time.

8.3 Objectives and Status

The following bullet items present the initial objectives for this Pilot Study and the current performance results for each objective, based on the progress made to date as described above:

1. Increase the moisture of the refuse.

   Based on temperature increases noted from the thermocouple points (see Tables), the moisture of the refuse appears to be increasing within a 75-foot radius around each injector. The moisture increase comes from steam moisture and bio-degradation of refuse, both related to the increase in refuse temperature.

2. Increase the temperature of the refuse.

   The temperature data also indicates that there are areas within the test site that have increased in temperature by as much as $50°F$. Additional information regarding temperature increases in specific areas will be analyzed during the upcoming PPT investigation that will be performed near the completion of the Study.

3. Monitor the migration of the steam through the refuse horizontally and vertically.

   The steam migration through the refuse, both horizontally and vertically, has been continually monitored during the Study period. Temperature increases have been detected at 15 feet below ground surfaces as well as 35 feet below ground surface.

4. Control the steam migration by using the LFG collectors.

It has been demonstrated that steam migration through the refuse can be controlled by increasing or decreasing the vacuum flow at the LFG collectors.

5. Monitor any excess liquid at the bottom of the test site.

Piezometer readings taken during the process do not indicate the presence of any liquid collecting beneath the refuse layer (50ft) at the contact with impervious fill cap soils.

6. Evaluate whether or not landfill leachate and condensate can be used in this process.

This has been the greatest challenge and appears to have been resolved. At this point, the leachate and condensate from the landfill can be used in this process assuming the landfill can provide a sufficient amount of leachate and the filtration operation can maintain production.

7. Increase the LFG quality and quantity output within the test site.

The primary goal of this Pilot Study is to control the flow of the collectors so that control of the migration of the steam can be maintained, and is not intended to obtain the highest flow rate of LFG. However, as a positive addition, some of the LFG collectors have indicated a 10% increase in methane concentrations. This appears promising, although each time the collector valves are closed to retain the heat generated because water inflow has ceased, either due to insufficient water provided or system down time for repairs/maintenance, the flow rates become impossible to accurately calculate. As production times become stable and longer term, more consistent measurements can be made. More definitive information will be available at the time of the next report.

8. Increase the settlement of the refuse

    There is visible evidence along the line of the steam injectors of about 18" to 24" of settlement. Soil settlement has not yet been detected at the settlement monuments; however, the Study was focused on the viability of the process and the potential environmental impact of moisture added to the refuse. The reduced volume of leachate water available for the Study has prevented settlement from spreading out far from the injection wells at this time. Additionally, the existing surface soil cap is likely bridging over the refuse. Significant settlement is anticipated when surface compaction efforts collapse the soils.

9. Obtain material quantities and costs per acre for treatment.

    A cost analysis will be presented after the study is concluded and the information reviewed and analyzed. Initially, the additional costs for larger boilers and filtration systems to overcome the high solids content of the leachate water, and, the recent substantial increase in fuel costs have had an impact on the initial cost estimates for the Study.

10. Obtain landfill surface settlement values during a 3 to 6 month period.

    Because of the filtering of the leachate water, the Study is about two months behind schedule. As long as there is insufficient water to convert to steam, the Study will not achieve its desired potential. Surface settlement values will not be accurate or significant if the Study cannot continue as planned with appropriate supply of water. Monitoring will continue and be reported at the completion of the Study.

8.4 Conclusions

As stated above, despite production issues and delays, most of the Study's objectives have been achieved.

With the modest amount of water that has been injected and the amount of settlement achieved, the progress is encouraging. Once there is enough water to operate 24-hours a day for several weeks, another PPT Profile will be performed to evaluate actual subsurface conditions relative to the anticipated results.

It is becoming apparent that the soil layers used to cap the landfill at various times, are bridging over the underlying refuse layers and preventing settlement despite the increase in void space within the refuse zones. Therefore, in order to fully realize the actual volume reduction potential, some form of mechanical compaction may be necessary to break through the soil layer and allow total settlement to occur. Evaluation of the impact of this condition will be conducted as production is increased and the system is operating at full expectation.

By converting the collectors EW-2 and EW-5 to injectors it confirmed that the collectors and injectors are interchangeable. Burying the steam hose under the cover soil verified that this is a good alternative to suspended steel pipes with insulation.

Chapter 9

In March 2006 the Pilot Study was terminated and this final report was submitted.   This is an abridged version to minimize duplication from the previous reports. The actual final report was a stand-alone document.

9.0 Introduction

This is the Final Report for the Steam Injection Pilot Study that was performed at the Miramar Landfill in San Diego, California. Field investigations began on May 16, 2005 with the PPT subsurface profiling and installation of the LFG collectors, injectors and monitoring instrumentation. On June 15, 2005 the steam boiler was brought on line and the steam process began by injecting 700 gallons of landfill leachate water on the first day. An Interim Report (Reference) was presented to the City, describing the investigation process and results, as well as outlining the locations and depths of the collector wells and injector wells. In November 2005, a second Interim Progress Report was submitted to the City and the agencies. Since that time the Pilot Study continued as described below until March 17, 2006 when the study was terminated.

9.1 Progress

Only about 350,000 gallons of water was injected as steam, which is well below the 800,000 gallons initially planned. The landfill was only producing an average of about 1,500 gallons of leachate and condensate per day, insufficient to achieve the planned objectives, whereas, the planned steam injection process requires 3,400 gallons per day, yielding a water shortage of 1,900 gallons per day.

A request was submitted to the City to use locally available recycled water as a supplement to the leachate and condensate water. This would allow for production goals to be achieved.

However, this request was rejected by the Regional Water Quality Control Board, due to the short term of the pilot study. The Board stated that the application would be considered if or when the project went to full scale.

Although there was limited water it was necessary to increase the size of the boiler to increase the steam influence in the test site. A 600,000 Btu unit was ordered but had a 1 month lead-time so a used 440,000 Btu unit was used until the larger unit arrived. The larger unit did increase the steam influence.

During the study, temperature readings were recorded, piezometer readings taken, and, LFG readings were measured using a Landtec Gem 2000. The chronological results of the various measurements have been tabulated and are included at the end of this report. Generally, all aspects show favorable increases during the monitoring period, despite the delays and inconsistent operation times.

9.2 Operations

The problems encountered due to the high concentrations of solid particles within the landfill leachate water being used in the injection system have resulted in significant cost increases and delays. Due to these delays the Pilot Study was about 2-months behind the planned schedule. The impact of this delay on the Study essentially required the continuation of operations into the rainy season. As a result, earthen berms were constructed around the test site to control surface water drainage. On a positive note this delay allowed us to operate through a full cycle of weather changes and the challenges presented to us.

Fortunately the weather was a typical rainy season with only a few heavy rain events. There were only a few days where the tanker trucks could not get to the site due to muddy conditions. No adverse impacts occurred to the landfill or the environment due to the Pilot Study operation.

One of the goals of the study was to evaluate the use of the push-in steel collectors and injectors with oilfield mill slot screens.

The purpose of using this system is to allow the collectors to be converted into injection wells if necessary and vise versa. Two of the smaller steam boilers were connected to EW-2 and EW-5, making those two collector wells into steam injection points. Steam was injected into these injectors for a short time but the lack of water minimized the length of this operation. However, the injectors did perform as planned proving that this system can be used as injectors or collectors. There was no evidence that steam ever leaked from around the injectors and no oxygen was indicated in the gas collectors. Burying the steam hose under the cover soil did insulate and protect the hose well.

On February 24, 2006 a 30-ton PPT rig was mobilized to the study site. A series of PPT soundings were performed adjacent to the first soundings across the control area and the test site. The findings are discussed later in this report.

On March 17, 2006 all operations were terminated and demobilizing of equipment and personnel began.

9.3 Objectives and Status

The following bullet items present the initial objectives for this Pilot Study and the performance results for each objective:

1. Increase the moisture of the refuse.

   Based on temperature increases noted from the thermocouple points (see Tables), the moisture of the refuse appears to be increasing within a 75-foot radius around each injector. The moisture increase comes from steam moisture and bio-degradation of refuse, both related to the increase in refuse temperature. This was substantiated by turning off the steam injection over a 24 to 36 hour interval and observe the thermocouple readings. The temperature readings indicated that the increased temperature of the refuse was self-sustaining due to biodegradation of the organic portion of the refuse by increased moisture content.

2. Increase the temperature of the refuse.

The temperature data also indicates that there are areas within the test site that have increased in temperature by as much as 50° F.

3. Monitor the migration of the steam through the refuse horizontally and vertically.

The steam migration through the refuse, both horizontally and vertically, was continually monitored during the Study period. Temperature increases have been detected at 15 feet below ground surfaces as well as 35 feet below ground surface.

4. Control the steam migration by using the LFG collectors.

It has been demonstrated that steam migration through the refuse can be controlled by increasing or decreasing the vacuum flow at the LFG collectors.

5. Monitor any excess liquid at the bottom of the test site.

Static Piezometer readings taken during the process did not indicate the presence of any liquid collecting beneath the refuse layer (50ft) at the contact with impervious fill interim cover soil layer. The post PPT Profile did not indicate any perched water layers in the waste prism.

6. Evaluate whether or not landfill leachate and condensate can be used in this process.

This has been the greatest challenge and appears to have been resolved. At this point, the leachate and condensate from the landfill can be used in this process assuming the landfill can provide a sufficient amount of leachate and the filtration operation can maintain production. This study also confirmed that most re-circulated leachate mostly flows through the waste with very little of it being retained. The landfill was discharging about 13,000 gallons of leachate per day at the beginning of the study. At the end of the study the landfill was only discharging 500 gallons of leachate per day. With the water being injected as steam instead of being re-circulated the daily discharge was greatly reduced. None of the injected steam ever reached the bottom drain. It is assumed it was converted to LFG.

7. <u>Increase the LFG quality and quantity output within the test site.</u>

The primary goal of this Pilot Study is to control the flow of the collectors so that control of the migration of the steam can be maintained, and is not intended to obtain the highest flow rate of LFG. However, as a positive addition, some of the LFG collectors have indicated a 12% increase in methane concentrations. This appears promising, although each time the collector valves are closed to retain the heat generated because water inflow has ceased, either due to insufficient water provided or system down time for repairs/maintenance, the flow rates become impossible to accurately calculate. During the summer months the methane concentrations in the collectors outside the test area dropped to 45% while the methane concentrations in the test area stayed at 66%. It was observed that at the start of the study little gas flowed from the condensate drain valves. Once the bio-process began, the gas flowed under high pressure from the condensate drain valves.

8. Increase the settlement of the refuse

There is visible evidence along the line of the steam injectors of about 18" to 24" of settlement (see photo below). However, the survey data obtained by the City surveyors indicate that only about 12" of settlement occurred at some of the points surveyed (see Settlement Table). The reduced volume of leachate water available for the Study prevented settlement from spreading out far from the injection wells. Additionally, the existing surface soil cap is likely bridging over the refuse. We are confident that with sufficient amount of water to allow for 24/7 operations the rate of settlement would have been much greater. To break the bridging of the cover soil it was hoped that the landfill operator would run one of their loaded scrappers or water tows over the site at the end of the study and before the post PPT Profile. However, the day before the PPT was to mobilize to the site the City refused to send their equipment to the study area for fear it would fall into a big hole. The 30-ton PPT rig did drive over the site and did not fall into a big hole.

9. Obtain material quantities and costs per acre for treatment.

The target cost for performing steam injection in landfills before the study was approximately $4,500.00 per acre. Naturally the cost of a Pilot Study was expected to cost much more due to the PPT studies, instrumentation and system design. With solar or geothermal pre-heating of the water, using landfill gas and not counting the cost of water, this target should be able to be achieved. Over time and as most of the equipment is reused for each acre treated the costs will drop dramatically.

116

10. Maintain Regulatory Compliance.

The Pilot Study was operated within regulatory guidelines. We complied with all ISO 14001 Standards and operated within the Environmental Management System. No apparent negative impacts were indicated to the environment or the landfill.

9.4 PPT Results

At the conclusion of the study, PPT soundings were performed close to the locations of the collectors, injectors and grid points. The PPT sounding logs of the first series at the start of the study were overlaid by the soundings of the last series of PPT soundings. They are not presented in this book because they would be too difficult to read at this size of these pages. Other soundings indicated a shift in elevation in some of the spikes compared between adjacent soundings. This is probably due to settlement between dense layers at the mid-depth of the test cell.

This PPT rig was used to perform the pre and post study investigations. The PPT rig did not have a large enough center hole to install the push-in collectors and injectors so a geoprobe type direct push unit was used. Most PPT contractors have rigs with larger center holes that will pass the 2" diameter pipes thereby saving costs with only one rig.

Figure 9-1   **Figure 9-2**

The PPT rig was used to install 9 thermocouple probes at various locations in the test area. The temperatures were displayed on a scanner presented in the photo on the right.

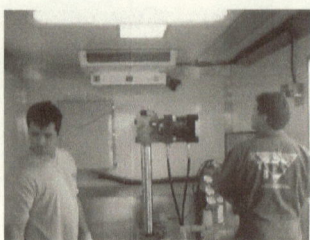

Figure 9-3

Figure 9-4

The photos below are of a completed gas collector and a completed steam injector on the right.

Figure 9-5

Figure 9-6

The study was started with the 360,000 Btu boiler on the left and increased to a 600,000 Btu on the right.

Figure 9-7

Figure 9-8

Static Piezometers (left) were installed in the test area to monitor for excess liquid at the bottom of the test cell. No liquid was ever indicated. Settlement monuments were installed to measure settlement at six locations.

Figure 9-9

Figure 9-10

Settlement

The steam injection pipeline was a good reference point to monitor settlement across the study site. As the ground lowered the pipeline had to periodically be re-supported along its length.

Figure 9-11

119

Figure 9-12

Figure 9-13

In the photo below, the wood stake is where the pipeline was supported when the test first started. The measuring tape shows over 24 inches of airspace.

Figure 9-14

120

Two months from the start of the study, stress cracks began to appear. As flags were placed in the cracks a pattern began to form as indicated in Figure 9-15. A circular line of cracks was formed in about a 75-foot radius around the injectors.

Figure 9-15   **Figure 9-16**

9.5 Findings

The following are the findings of the Pilot Study.

- The refuse in the study cell was very dry preventing the entire void space to be filled with steam. The refuse absorbed most of the moisture as it was injected unless the vacuum was increased to a level that caused the steam to migrate though the waste prism. The lack of water prevented a prolonged injection surge to overcome the dryness and reach the full 100-foot radius of the study cell.

- As the ambient temperature at the study site cooled, the water-cooled and propane usage increased by about 5%. Pre-heating the water by using solar panels or geothermal wells, as much as 50% could be saved on fuel costs. For colder regions STI is developing a geothermal well to be installed in landfills using the warm waste heat to pre-heat the water. This could save as much as 30% on fuel costs. For optimal savings and efficiency both should be used in tandem.

- An average of 3 inches of settlement per month was measured at some of the survey points on the study site. However, other observations around the site indicated that the settlement was not uniform across the site (see photos). It is also believed that the thick cover soil, as much as 8 feet, may be inhibiting the rate of surface settlement by bridging. The PPT logs indicate low-density layers under the cover soil, which may indicate that a large amount of organic waste has biodegraded.

- At the start of the study the methane concentration was 54% in the test site collectors. Within a couple of months the methane concentration increased to about 66%. During the summer months the existing collectors outside the study area were extracting moisture from the waste prism as they were collecting landfill gas. During this time the methane concentration decreased to 45%. At this same time as steam was replenishing the moisture in the study area the methane concentration remained at 66%.

- Installing and maintaining the steam pipeline was labor intensive and costly. To improve future operations a rubber steam hose should be used instead. Birds constantly tore off the insulation, which required constant repairing. To insulate and protect the hose it should be buried under the cover soil. This will also reduce corrosion.

- The collectors and injectors performed as planned and they are inter-changeable, which may be useful during full scale areas along the edges of the landfill that may need to have steam injected in small pockets of refuse that were not affected during the main injection process.

- Thermocouples were effective in detecting the migrating steam. Occasionally where the wires were spliced together the contacts would corrode and would be shorted. Once they were repaired they operated as expected.

9.6 Conclusions

As stated above, despite production issues and delays, most of the Study's objectives were achieved. With the modest amount of water that was injected and the amount of settlement achieved, the process is encouraging. With enough water to operate 24-hours a day settlement rates should increase dramatically.

It is becoming apparent that the soil layers used to cap the landfill at various times, are bridging over the underlying refuse layers and preventing settlement despite the increase in void space within the refuse zones. Therefore, in order to fully realize the actual volume reduction potential, some form of mechanical compaction may be necessary to break through the soil layer and allow total settlement to occur.

Temperature Table

Location	Probe Depth	Start Temperature	End Temperature	Total Increase
	Feet	June 7, 2005	March 15, 2006	Degree F.
T-1	22	97	116	19
T-2	22	94	95	1
T-3	22	96	97	1
T-4	22	93	95	2
T-5	31	96	126	30
T-6	15	88	157	69
T-7	35	95	100	5
T-8	15	86	98	12
T-9	31	96	108	12

Settlement Table

Location	Start Elevation	End Elevation	Total Settlement
	May 25, 2005	March 15, 2006	Feet
SM-1	449.8	449.4	0.4
SM-2	446.7	446.3	0.4
SM-3	445.1	444.9	0.8
SM-4	446.5	446.2	0.3
SM-5	449.0	448.9	0.1
SM-6	450.3	450.1	0.2
SI-1	449.3	448.0	1.3
SI-2	447.4	446.3	1.1
SI-3	444.7	444.0	0.7
EW-1	448.4	448.4	0.0
EW-2	446.4	446.3	0.1
EW-3	444.2	444.2	0.0
EW-4	446.2	446.0	0.2
EW-5	449.0	448.5	0.5
EW-6	450.5	450.3	0.2
CEW-1	451.7	451.6	0.1
CEW-2	448.6	448.5	0.1
PPT-10	447.9	447.5	0.4
PPT-12	449.2	448.8	0.4
PPT-18	447.5	447.4	0.1
PPT-19	450.0	449.6	0.4

PPT Summary Table
Summary of PPT Soundings and EW, Thermocouples Installations

PPT Number	Percentage Of Refuse	Percentage Of Soil	Depth Of Screen	Total PPT Depth
	(50 ft column)	(50 ft column)	(feet bgs)	(feet bgs)
PPT-8	70	30	0	68
PPT-10	60	40	0	54
PPT-10-2	75	25	0	50
PPT-11	50	50	0	53
PPT-12	60	40	0	54
PPT-12-2	75	25	0	50
PPT-13	70	30	0	50
PPT-17	70	30	0	52
PPT-17-2	70	30	0	50
PPT-18	60	40	0	50
PPT-18-2	60	40	0	50
PPT-19	70	30	0	48
PPT-19-2	50	50	0	50
PPT-20-2	60	40	0	50
Collectors				
EW-1/PPT-8	70	30	20 to 35	Repeat
EW-1-2	50	50	0	53
EW-2	60	40	25 to 40	53
EW-2-2	50	50	0	50
EW-3	55	45	15 to 30	45
EW-3-2	50	50	0	50

Summary Table
(continued)

PPT Number	Percentage Of Refuse	Percentage Of Soil	Depth Of Screen	Total PPT Depth
Colletors	(50 ft column)	(50 ft column)	(feet bgs)	(feet bgs)
EW-4	70	30	20 to 35	74
EW-4-2	70	30	0	50
EW-5/PPT-13	70	30	20 to 35	Repeat
EW-5-2	50	50	0	50
EW-6	70	30	20 to 35	60
EW-6-2	70	30	0	50
CEW-1	50	50	20 to 35	50
CEW-1-2	50	50	0	50
CEW-2	70	30	17 to 37	60
Cew-2-2	70	30	0	50
Steam Injectors				
SI-1	70	30	30 to 45	55
SI-1-2	70	30	0	50
SI-2	60	40	20 to 35	56
SI-2-2	60	40	0	50
SI-3	70	30	20 to 35	50
SI-3-2	60	40	0	50
Thermocoup les				
T-1				22
T-2				22
T-3				22

Summary Table
(continued)

PPT Number	Percentage Of Refuse	Percentage Of Soil	Depth Of Screen	Total PPT Depth
	(50 ft column)	(50 ft column)	(feet bgs)	(feet bgs)
T-4				22
T-5				31
T-6				15
T-7				35
T-8				15
T-9				31
T-10 – SI-1				Top of Injector
T-11 – SI-2				Top of Injector
T-12 – SI-3				Top of Injector
Total Footage			**180**	

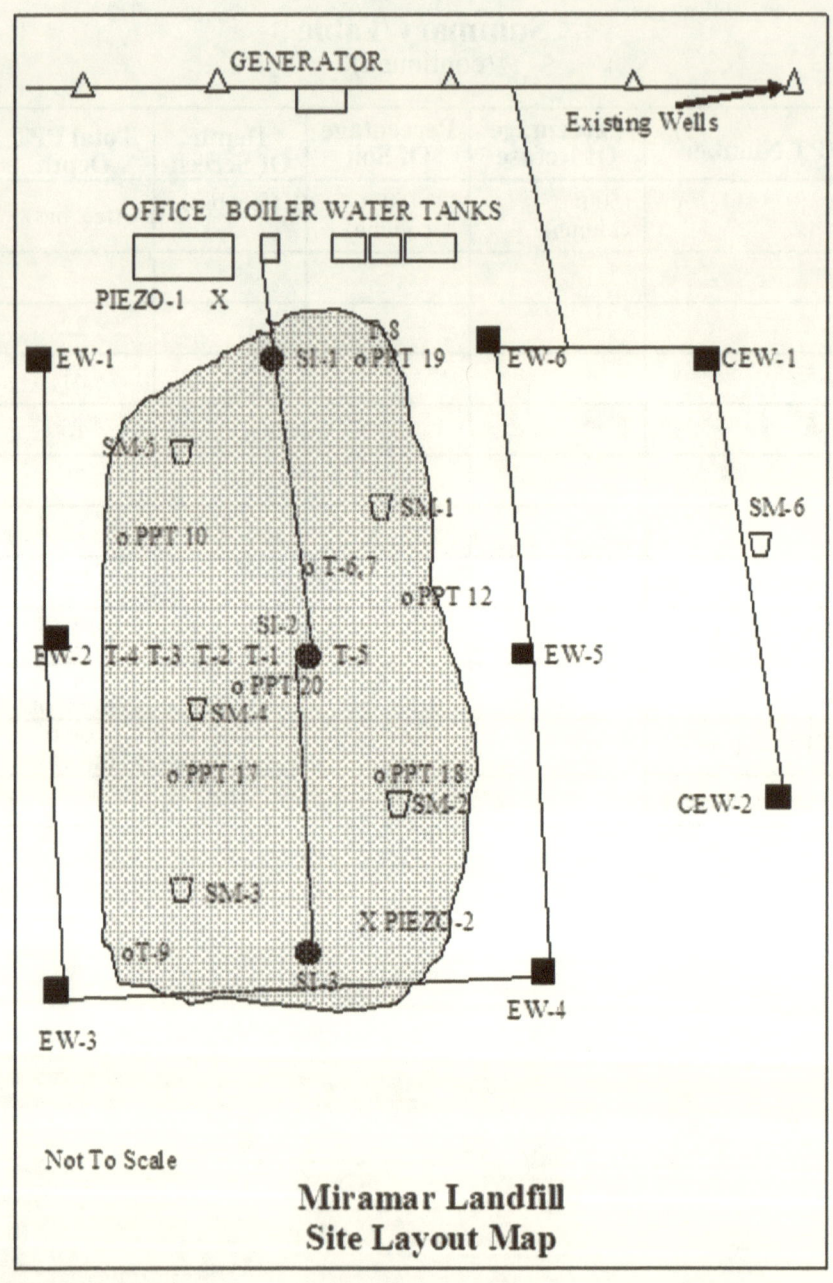

**Miramar Landfill
Site Layout Map**

Not To Scale

Figure 9-17

128

Chapter 10

Although the main purpose of the Pilot Study performed at Miramar Landfill was to verify the effectiveness of steam injection to recover airspace the other purpose for steam injection is to enhance landfill gas generation for operating a power plant. The following chapter outlines the landfill gas generation potential using the Steam Injection Process. As you will see the actual energy potential stored inside a municipal sanitary landfill is incredible and is being wasted with dry tomb landfills.

10.0 Enhanced Landfill Gas Generation

The three things that can diminish landfill gas generation are to allow air into the landfill, cool the waste and to remove moisture. As soon as you begin to extract LFG, water is being removed from the waste prism since LFG is saturated with moisture. If the moisture is not replaced the gas generation rate will diminish. As the gas generation rate slows the waste prism will cool. The only way to enhance the generation of landfill gas beyond natural processes is to replace the moisture and heat that's been removed by extraction.

The following calculations are based on basic physics, published data and past experience. A past Pilot Study using steam injection in Southern California showed that no water accumulation occurred in the waste prism. It is assumed that most of the steam was converted by anaerobic bacteria into LFG during the study. For this outline it is assumed that 5,000 gallons of water per day, heated to high-pressure hot water and is converted to steam inside the waste prism. The following demonstrates the gas potential of an acre of waste 75 feet deep. Actual recovery rates may vary due to variable conditions in the landfill.

1 acre 208 x 208 feet x 75 feet deep = 3,244,800 cubic feet of refuse x .20 percent void space = 648,960 cubic feet. 3,000 gallons converted to steam would fill the void spaces.

Five thousand gallons will overcome the dry conditions of the waste and other losses in the system.

5,000 gallons / 7.5 gal. per cubic foot = 666.7 cubic feet X 1,600 expansion factor of steam = 1,066,720 cubic feet of steam. The steam vapor and heat are the mechanisms that enhances the anaerobic conversions occurring in a landfill to produce LFG. It is known that the organic material in the landfill contains the necessary components (hydrogen and carbon) to produce methane. However, the additional heat and moisture accelerates this process. Based on the airspace recovered and the increase in gas volumes during our Pilot Study it is assumed that one volume of steam will facilitate creating 1/2 volume of LFG. 1,066,720 cu. ft. of steam will create 533,340 cubic feet of LFG, which contains .50 percent methane = **266,703 cu. ft. CH_4/day**. Although we actually produced 62% to 66% methane in our Pilot Study in San Diego, CA.

If 1 cubic foot of methane produces 1012 BTU (Wikipedia) and a typical engine will generate power at 10,500 Btus/kwhr. Then 266,680 cu. ft. CH_4 X 1012 BTU / 10,500 Btus/kwhr has the potential to produce **25,703 kwhrs per day per acre**.

There could be a 3 to 4 day delay until the dry conditions of the refuse are overcome and the bioreaction is sustainable. As the bioreaction increases so does the temperature so the steam injection can be reduced or diverted to another acre until the bioreaction slows and then the steam would be reapplied.

10.1 Organic Potential

1 ton or 3.5 feet cubed or (42.9 cu. ft. or 47 pcf) of waste with 50% organic material will produce about 6,000 cubic feet of methane gas (MSW, Neal Bolton).

1 acre - 75 feet deep of fresh waste will contain an estimated 1,622,400 cu. ft. of organic material.

There is an estimated potential of **227,000,000 cu. ft. of CH_4** or **22,260,000 kwhr / acre.**

There is also an industry compaction standard of 52 pounds per cubic foot, which would increase the estimated potential to **247 million cu. ft. of CH₄** or **23,500,000 kwhr per acre.**

With the above production rate the gas production should last **22,260,000 kwhr / 25,703 kwhr / day = 866 days or 2.4 years.** Or with 52 pcf = **1,023 days or 2.8 years.**

These conclusions are based on injecting the 3,000 to 5,000 gallons of water per day in a perfect world however, some loss to very dry waste will occur. Based on natural decomposition, the same 1-acre will take 100 years or more to stabilize. But it will never deliver the same LFG in a productive rate that could be used for the last 50 years.

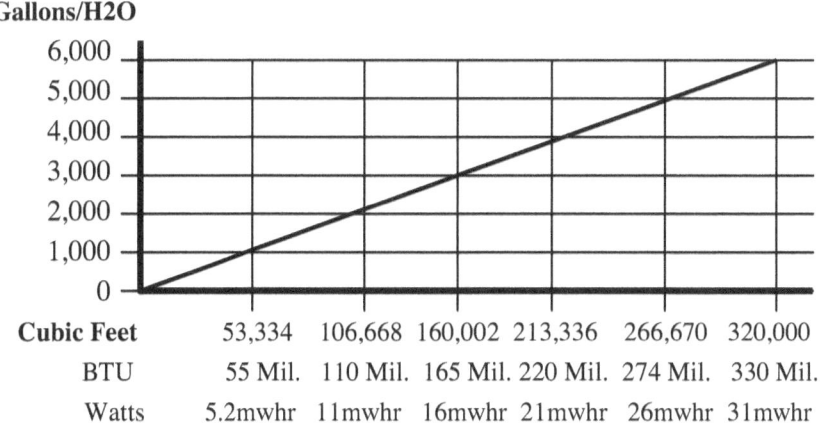

Methane Production
For 1 Day and 1 Acre

Figure 10-1

Some power companies claim that their new generating engines are now closer to 60% efficiency than the earlier 30% efficiency. If this is the case instead of 25,703kwhr/day it would be 52,324kwhr.

With more aggressive recycling efforts, with the removal of inorganic material, the organic content of the waste prism should increase in the future, making this process even more productive.

10.2 Energy Use

To enhance the generation of LFG it may require the need to consume energy. As in any fuel recovery system it is necessary to consume less energy than what is recovered to be successful.

The most cost effective means of heating the water to 250°F is to use the waste heat from the power generating system. This should be attempted first, providing 100% of the methane being generated for power production.

However, depending on the layout of the landfill's collection system and the location of the power plant, piping the steam to the opposite side of the landfill may not be practical, especially in cold regions. It may be more practical to place a small portable boiler closer to the treatment area. Some of the LFG (about 10 scfm at 50% methane) would be pulled from the local header system to fire the boiler.

To conserve energy the water should be preheated prior to entering the boiler.

- A solar collector can increase the water temperature to 120°F+. It will require 5,395,000 BTU for an increase of 130°F to heat the water to 250°F. For example: 130 degrees X (5,000 gallons X 8.3 lbs. per gal. = 41,500 lbs.) = 5,395,000 BTU or **5,238 cubic feet of methane per day.** An insulated tank can be used to store more warm water during the day and used for nighttime use.

- In cold regions the water can be circulated through a geothermal well installed into the waste prism, which would heat the water to about 85°F to 90°F. When the steam begins the bioreaction process, the water would then be heated to 120°F as well.

For more uniform day/night pre-heating, both methods should be used piggybacked.

132

The worst case scenario would require 5,238 cubic feet of methane, which is **less than 2.0 %** of the potential gas generation per day, leaving 98 % for power generation.

Steam injection puts control of LFG production in human hands. Only put in as much steam as required to operate the power plant at peak power.

Energy Consumption
For 1 Day and 1 Acre

To conserve energy the water should be preheated by solar, geothermal, or heat exchanger prior to entering the boiler. The following graph assumes the water has been pre-heated to 120°F, the goal is to heat the water to 250°F.

Gallons/H2O

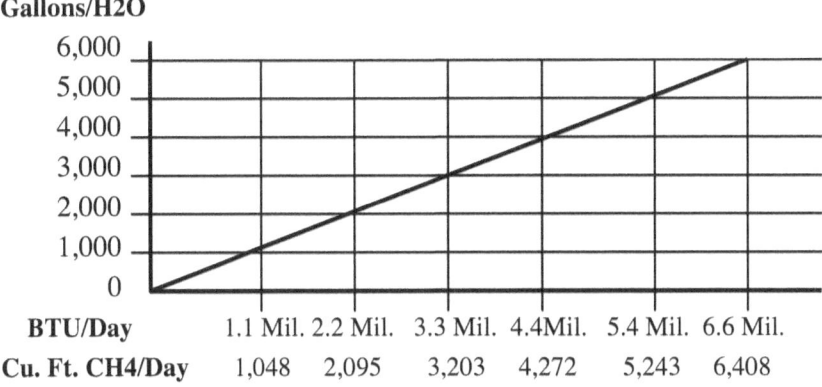

BTU/Day	1.1 Mil.	2.2 Mil.	3.3 Mil.	4.4Mil.	5.4 Mil.	6.6 Mil.
Cu. Ft. CH4/Day	1,048	2,095	3,203	4,272	5,243	6,408

Figure 10-2

Less than 2.0 % of the fuel recovered is used to heat the water in a heating unit if the heat is not recovered from the power plant.

Upon review of the gas flow readings during the San Diego Pilot Study, data indicates that the organic conversion rate was a 1:1 conversion ratio. This means for every volume of steam injected 1 volume of LFG was produced.

Chapter 11

In 2003 I received a call from Nassau, Bahamas requesting that I come down and look at their burning C&D landfill and see if steam injection could put it out. So I flew down and had a few nice days in Nassau except for the part about visiting the burning landfill. It looked like something from Dante's Inferno and smelled worse. When I got back to my hotel room I had to put my cloths out on the patio. The landfill was a lost cause they ended up de-constructing the whole cell and starting over using modern methods of compaction and covering.

11.0 Subterranean Landfill Fires

Landfill fires are some of the most elusive conditions to identify and deal with. They can create sink holes, emit smoke, degrade the gas quality and damage bottom liners.

11.1 The Cause

The three basic elements of fire are fuel, heat and oxygen. All of these elements can be found at all landfills. The most obvious way of preventing subterranean fires is to control at least one of the elements of fire. Usually, this is done by preventing oxygen or air from entering the trash prism of the landfill. However, this is not always possible, the following are some of the ways air can enter a landfill:

- Air is trapped in the refuse as it is placed into the landfill (starved air combustion).
- Insufficient soil cover over the refuse.
- Poorly located, designed, or constructed LFG collectors.
- Overdraw on LFG collectors.
- Buried haul roads (particularly gravel roads) provide pathways for air intrusion.
- Separation of the bench road from the slope.

Often, if a large amount of refuse is placed in warm weather and not compacted properly rapid composting can occur and the refuse will begin to smoke. If the refuse is not covered with sufficient soil to prevent air infiltration, subterranean fires can be started. This can happen if insufficient soil is not used in the first place or soil erosion has removed the cover soil and allowed air to infiltrate the landfill.

Once the refuse is in place and covered with soil, usually LFG collectors are installed into the landfill to extract LFG. During this process, it is important that the collectors are designed, located and constructed properly. If the vacuum is adjusted too high, at a particular collector, it will overdraw the area of influence and air will enter the refuse either from the top or through an adjacent slope of the landfill. As air enters the refuse, subterranean fires can occur.

As a landfill increases in height a haul road that was used to transport refuse to the working face is often buried with refuse. The edges of these haul roads are not compacted like the center of the road where traffic had flowed. When the roads are buried these soft edges can become a pipeline to the slope of the landfill and provide a pathway for air into the landfill and for LFG to escape to the atmosphere, particularly if gravel is placed on the roads.

It has been observed that over time a bench road can separate from the side of the landfill as settlement of the landfill occurs. When this happens, air and rainwater can enter the refuse and start subterranean fires if the fissures are not appropriately repaired.

11.2 Troubleshooting the Problem

Currently, there are four basic indicators that will allow an operator to know if a subterranean fire is present at a landfill.

- Smoke emitting from the landfill.
- Elevated gas temperatures in LFG collectors or header.
- Rapid settlement or sinkholes usually around a collector.

- Elevated carbon dioxide (CO_2) and/or carbon monoxide (CO) levels in gas stream.

Troubleshooting a subterranean fire or rapid decomposition can be easy if smoke is emitting from the landfill and the cause can probably be focused on a nearby collector that is overdrawing. However, it is not always that easy. If the gas temperature in a particular collector is elevated, (140 + degrees F) but no smoke or sinkholes are not visible, it can be difficult to determine where the subterranean fire is located.

Currently, most operators have to just monitor the situation until it becomes visibly obvious where the fire is located by smoke or sinkholes. This may be too late to avoid damage if there is a bottom liner in place at a particular landfill. Water quality agencies are very concerned about subterranean fires when there is a bottom liner involved.

The alternative is to perform a Piezo-Penetrometer Test (PPT) and using a thermister to measure the temperatures as the PPT cone is pushed into the landfill. The PPT can provide the following troubleshooting abilities:

- The PPT can map out possible void spaces in the refuse.
- The PPT can locate layers of vacuum, which may be pulling in air.
- The PPT can locate buried haul roads.
- Temperature readings will focus in on the affected area.
- In-situ gas samples can be recovered and tested for CO_2.

A grid pattern of PPT soundings would be performed in the area of concern. The PPT would also profile the vacuum influence of nearby collectors and detect possible voids from low tip resistance that may become sinkholes. Primarily, due to the PPT's accurate depth control, it will be able to indicate the temperature near the bottom liner without punching through it as long as accurate surface elevation is provided prior to the test.

The PPT would also be used to locate possible piping caused by buried haul roads or from leaking bulkheads from horizontal collectors. In-situ gas samples can also be recovered using a down-hole gas sampler to check CO_2 levels.

Once the PPT grid is complete, the digital data would be used to create a 3-D Profile of the area of concern. With this information, a plan of action can be developed to extinguish the subterranean fire.

11.3 Extinguishing Subterranean Fires

Obviously, if the landfill is smoking around a particular collector it is probably due to overdrawing and the vacuum control valve on the collector should be restricted to reduce flow. Once this is done, smoke may start emitting from the place where the air was entering the landfill that caused the subterranean fire. It may be necessary to add more soil in this area and track-roll it in. Oxygen starvation is one way to retard combustion but it may not correct the reason it started in the first place. If the operator increased the vacuum influence of a particular collector, it was probably because there was a gas emission problem in the area. Once overdraw began, it was not only pulling in air, but it was drying out the refuse to a point where it could burn or begin rapid composting.

By just using air starvation techniques it will stop the fire but the gas emission problem will also return and if the same collector is opened again to respond to this condition the fire will also return.

This scenario is very common and the best way to solve this problem is to install another collector closer to the surface emission area. However, it is not always obvious as to the real source of the LFG production especially if the surface emission is on a slope. The PPT should be used to pin point the gas layer that is migrating to the slope and a Push-in 2" diameter steel collector can be installed with screen sections just at the gas layer. This approach will solve the overdraw causing the fire and still provide gas emission control.

As mentioned in the previous section, if the PPT Profile should indicate that a large area has been impacted by a subterranean fire, a more aggressive approach may be required to extinguish the fire. Currently, nitrogen/CO_2 injection is used to cool the refuse and displace the air that may be feeding the fire. Another alternative is to inject steam into the landfill.

Steam has many advantages over nitrogen/CO_2 injection but the most obvious one is that it will not only put the fires out but it will change the conditions that caused the fires. By replacing the moisture in the refuse, there is less chance of the fire returning, nitrogen or CO_2 will only dry out the refuse more. Steam will also increase LFG production and not dilute the gas stream as with nitrogen/ CO_2. By using the PPT rig to push-in the steel steam injectors and collectors, there are also cost benefits with steam injection costing about $2,000 to $3,000 less per acre using steam instead of nitrogen and $1,000 to $2,000 less than CO_2.

Once the subterranean fire is under control, the steam can continue to be injected into the landfill to increase LFG production by enhancing decomposition and increase airspace recovery. The steam injectors can also be used as LFG collectors in the future.

11.4 Prevention

Subterranean fires are a result of cause and effect of actions performed by the operator of a landfill. It is generally agreed that LFG collectors are the main cause of fires in landfills. Something usually occurs at a landfill (i.e. gas emissions, or subsurface migration, etc.) to require the operator to install LFG collectors in order to stay in compliance with regulations. As stated earlier in this book the current procedure is to blindly drill a large borehole sometimes as large as 2 feet in diameter to the full depth of the landfill at a location that is more convenient than it is scientific. The borehole is then filled with gravel or sand around a 4" to 6" diameter pipe. This can cause two problems right from the start.

138

The gravel column will permit any liquid that is keeping the refuse moist to drain to the bottom of the landfill and provides a short cut for air to infiltrate into the landfill.

As stated before Push-in collectors do not produce cuttings and they have no gravel and will not short-circuit to the surface.

If surface emissions of LFG should occur in the general area of a drilled in collector the first thing an operator is going to do is increase the vacuum influence on that collector to see if that will control the gas emission.

Sometimes this will work if the collector is not too close to a slope, but most times the surface emission occurs on a slope. If increasing the vacuum influence does control the gas emission, this will now indicate a connection between the air outside of the slope, to the gas layer and to the collector. Now instead of gas emitting from the slope, air is entering the landfill and if this continues over a period of time, the refuse will dry out and a subterranean fire will start. Often increasing the vacuum influence on the nearby collector does not affect the gas emissions because there is no vacuum connection to the gas layer that is migrating to the slope in the first place.

When the operator increases the vacuum influence the vacuum will find its own pathway to the slope, dry out the refuse, pull in air and start a fire. The only way to prevent this from occurring is to perform a PPT Profile in the area of concern, understand what is causing the surface emission and then take appropriate action to correct the problem. To blindly drill another collector can only increase the possibility of a fire by draining liquids from the upper portion of the landfill and introduce more air into the landfill.

11.5 Conclusions

The best way to prevent subterranean fires in landfills is by not doing the things that can cause them in the first place.

- Do not allow the soil cover over the refuse to deteriorate.

- Do not increase vacuum influence on collectors near a slope without understanding the cause of the surface emission. Perform a PPT Profile first.
- Do not blindly drill large diameter boreholes though the landfill. Pre-qualify the location with the PPT and install push-in 2" diameter steel collectors.

If a subterranean fire is suspected, it should be investigated with the PPT and if necessary, it should be extinguished with steam instead of nitrogen or CO_2.

Chapter 12

In 1999 I was conducting some experiments in my lab on a steam injector I was planning to use in the field. I was trying to determine if the steam would come out more at the top of slotted section than at the bottom. So I placed a 2" diameter 5' long steel injector inside a 6" diameter clear PVC pipe and filled the annulus with household waste. The injector was connected via steel piping to a small electric boiler used for steam presses in cleaners for pressing pants. Holes ¼" in diameter were drilled near the top, middle and bottom of the PVC pipe to allow for a thermocouple probe to be inserted. I set up video and still cameras and I was working alone. I started the boiler and the cameras and then started taking temperature readings along the column. It got real busy when I started taking still photos and temperature readings and checking the boiler valve settings. Before I knew it, the whole PVC column just melted and collapsed. This was a surprise since this was ¼" wall pipe. It was also a revelation as to the useful application of steam on MSW. I asked myself what would happen if you melted all the plastic in the waste stream going into a landfill and compact it before you placed it into the landfill. I started conducting Bench Sale Tests in my lab to find out.

By the way, it turns out that the steam travels down the injector pipe taking the path of least resistance and then out of the pipe into the refuse searching for another the path of least resistance.

12.0 Steam Injected Compaction Station

Baled waste operations are already being used at some landfills in Florida and New Jersey with some success. However, their compaction efforts are minimal due to the dryness of the waste and the elasticity of the plastics. It's like wading up a piece of paper and setting it on a table, it will unfold itself overtime. It usually takes plastic shrink-wrap or baling wire to hold the bales together adding more waste the landfill.

However, if you wet the paper and ball it up it will stay together. Also if you melt the plastic component of the waste, compact the waste and then cool the waste, the plastic component will hold the block together.

12.1 Airspace

Airspace – The most precious commodity found at a landfill second only to the refuse itself. The amount of airspace permitted to a licensed landfill will dictate the potential revenue that will be generated over its lifetime. Currently, landfill operators compact as much refuse into the allotted space as possible for maximum revenue. This process can cause side effects that are very costly and can diminish the amount of profit that could have been made. Typically, very heavy and costly equipment is used to place and compact the refuse.

When the refuse is densified it lowers the permeability of the waste prism (especially with high plastic content) and lowers the effectiveness of LFG collectors requiring closer spacing. Another cost that affects profit is daily cover and traffic decks. The soil used in this process does not go over a scale and collect tipping fees but uses as much as 1/4 to 1/3 of the airspace and does not decompose.

The best approach is to pre-treat the refuse by placing it into a compaction chamber and apply high temperature steam while compressing the refuse into a very compact block. This process will moisture condition and breakdown the cell structure of the organic waste but will also melt and shrink the plastic component of the refuse. The following report of a Bench Scale Test (BST) will demonstrate the effectiveness of this new approach.

12.2 Bench Scale Test

On September 17, 2003, I performed a bench scale test in the STI laboratory. The BST was performed using a stainless steel compaction chamber inside a load frame with an Enerpac hydraulic ram. The chamber was connected to a small electric boiler capable of producing +400° F steam.

Several tests were performed to demonstrate the differences between conventional compaction methods, steam applied methods and steam with confined compaction methods. Also, various ratios of organic material to plastic material were tested.

Constants

The following are the dimensions and constant values used in the calculations during the BST:

Chamber Inside Diameter – 2.0 inches = 3.14 sq. in.
200 grams = 0.44 pounds
Landfill Compaction - 4.76 inches @ 150 psi = 52 pcf =
1400 pounds per cu. yd.
Packer Truck Volume = 30 cu.yds./10 tons

The waste mixture used in these tests consisted of basic household waste such as plastic bags, newspaper, cardboard, compost, pieces of wood, Styrofoam, green waste and styrene.

Test No. 1

To observe the differences between the current method and the proposed method, it is necessary to determine what ram pressure to use to simulate the compaction effort applied to waste in the field.

This was accomplished by calculating the volume within the compaction chamber and the density of 200 grams of typical waste. A mark was placed on the chamber were the sample would be 4.76 inches long when compacted in its dry condition.

143

At this length the 200-gram sample would have a density of **52 pounds per cubic foot** (pcf) or 1400 pounds per cu.yd., which is usually found in many landfills. When the sample was compacted to 4.76 inches the pressure gauge on the hydraulic ram indicated 150-psi.

Test No. 2

The purpose of this test is to determine how much denser the refuse would become with the use of steam while maintaining the same compaction effort at 150-psi ram pressure.

Once the sample was compressed to 4.76 inches, 150 psi was maintained and 410° F steam was injected into the chamber. A thermocouple attached to the chamber indicated the temperature reached about 250° F on the outside of the chamber. At this time the ram pressure gauge indicated 200 psi due to the pressure of the injected steam. The steam was turned off and the ram pressure returned to 150 psi. The steam was reapplied and the ram was advanced while maintaining 200 psi. The ram advanced approximately 2.0 inches when the pressure began to increase so the ram was stopped. The steam injection was turned off and the sample was allowed to cool down below 100° F. The sample was removed from the chamber, measured and was found to be 3.0 inches long. This calculates that the sample has a density of **78.87** pcf or 2130 pounds per cu.yd. This equates to a **52%** increase in density.

Test No. 3

This test was a repeat of Test No. 2 but with higher ram pressures. As steam was applied the ram was advanced until the pressure gauge read 750 psi. At this time, steam stopped flowing through the waste and the pressure was rising in the boiler. This indicates that the density of the sample has affected the permeability of the waste. At this time the chamber temperature had reach 350° F. The ram advance was stopped and the sample was allowed to cool.

144

The sample was removed from the chamber and measured. The sample length was 3.0 inches in the chamber but due to the organic material mixed in with the plastic the sample rebounded when removed from the chamber. The sample expanded to 3.25 inches when removed from the chamber.

Several samples were tested with various ratios of organic material versus plastic (50:50, 70org:30p, 30org:70p) with little difference in density. The rebound seemed to be affected by the amount of organic material in the sample and the plastic was more uniformly melted when the chamber temperature was at 350 ° F.

It is believed that the limitation on increasing the density of these samples is due to shortcomings in the test chamber such as plastic plugging the steam port as the ram pressure was increased. This can be corrected in the full-scale model.

12.3 Packer Truck Compaction vs. Steam Injected Compaction Station

For this comparison the dimensions being used are an average of the Packer trucks found in the industry. On average these trucks contain about **30 cubic yards** of dry slightly compacted refuse and the waste weighs about **10 tons**. The density of the refuse calculates to be about **25 pcf** in a fully loaded truck.

When compared with the density from the compaction station at 79 pcf, this is a **216%** increase in density. If 4 truck loads totaling **120 cu. yds.** were compacted to 79 pcf, in a compaction station, the volume of the refuse would be reduced to a **26 cu. yd.** block or the volume of one truck load. This 4:1 ratio will save a significant amount of airspace in the landfill. This would also be very cost effective for transfer stations.

12.4 Conclusions

The BST has demonstrated that the use of a Compaction Station can significantly reduce the volume of waste going into a landfill.

By minimizing the volume of waste it will extend the life of the landfill and reduce the costs of operation, increasing profits. It will also reduce the amount of daily cover and traffic decks saving airspace for refuse. By eliminating most of the heavy equipment with their O&M costs, minimizing cover soil and recovering 52% + of the airspace, the use of Compaction Stations could double the profits for landfill operators. Also, with the uniform moisture conditioning of the organic waste this portion will decompose much faster and provide more reusable airspace.

Chapter 13

Well, the bench scale test was pretty impressive to me so I built a one cubic yard trailer mounted steam injected compaction station for about $200,000. I applied for a patent in 2000 and I'm still waiting for it to be awarded.

Figure 13-1

13.0 Steam Injected Compaction Station

Bioreactors have been used to reduce the organic portion of the waste in municipal solid waste landfills. The purpose of this is to produce more methane gas for co-generation operations and to recover airspace. This process is only useful after a significant amount of refuse has been placed in the landfill. The process can interfere with the day to day operations and does not reduce the amount of equipment needed to place the refuse in the landfill. To reduce the amount of equipment (dozers, compactors, dirt haulers, etc.) required to place the waste in the landfill, it is necessary to reduce the volume of the refuse going into the landfill.

To reduce the volume of the waste and to utilize the permitted airspace to its maximum capacity it is necessary to pre-treat the refuse prior to placing the waste into the landfill cell.

By pre-treating the waste with high temperature and high-pressure steam the volume of the plastic portion of the refuse will be melted and shrunk to its smallest volume. The organic portion of the refuse will be broken down and moisturized to promote rapid degradation. The driest portion of the waste, paper will be moisturized and make it more compactable. During this process the refuse is compacted into blocks and then they are transported to the landfill cell.

The following is a short outline of the proposed operation that is patent pending:

- Three to five standard garbage packer trucks would dump their loads onto a platform adjacent to the opening of the compaction chamber. A larger platform is used for transfer trucks. Several stations would be constructed to meet the daily tonnage rate. Two Compaction Stations can process about 2,000 tons per day.
- The refuse would be pushed by a loader or conveyed by a moving floor into the hopper of the compaction chamber (Figure 13-2).
- A hydraulic ram pushes the refuse into the chamber until the chamber is full but not compressed.
- The chamber is then sealed and high temp./high pressure steam is injected into the chamber. The ambient temperature of the refuse will dictate how long the steam will be applied approximately 5 to 10 minutes in California. Frozen refuse would be somewhat longer. As the temperature rises in the chamber the plastic component of the refuse will begin to melt and shrink with no off gassing.
- Once the refuse has reached the desired temperature (~300 to 500 degrees F) the steam would be extracted by applying a vacuum to the chamber. As long as the refuse stays above 250° F any free liquids in the refuse will turn to steam and be extracted and converted to water in the condenser unit and then pumped to the boiler.

- As the steam is being extracted the hydraulic ram is advanced into the chamber to compact the refuse into the smallest block possible based on the data provided during the scanning process.
- The block of refuse and the chamber would be quick cooled by water jackets in the walls of the chamber to about 200 $^\circ$ F. This will harden the plastic in the block and to prevent the next load from reacting prematurely to the heat and cause excessive vaporizing from the refuse while the chamber is open.
- Once the block is cooled, the chamber is opened and the ram continues to push until the block is clear of the chamber.
- The block which has reduced the volume of up to 4 truck loads of refuse to the size of 1 truck load, is then transported to the landfill cell and covered with a tarp.
- Once a layer of blocks have been placed in the cell, steel steam injector pipes are placed on top of these blocks horizontally.
- When another layer of blocks are placed on top of the first blocks, horizontal steel gas collectors are placed over these blocks.
- Once a cell has been filled to about 40 feet in height and covered in soil, steam would be injected which will open up the blocks and allow moisture into the blocks and biodegradation would begin. The gas collectors would extract the gas. The injectors and collectors can be used either way to open all the blocks and recover all the gas.

The following are the advantages of this process:

- Maximum use of airspace.
- Minimum amount of waste-handling equipment, maintenance and fuel.
- Reduction in site personnel.
- Garbage truck drivers always know where to go.
- Minimum amount of cover soil used, saving airspace and costs. No extended shifts for cover placement or removal.
- Rapid biodegradation of organic waste, airspace recovered.

- Refuse is more closely monitored.
- Less impact on environment less monitoring costs.

By reducing the volume of the waste prior to placing it into the cell, much less cover soil is required, which is one of the most costly tasks in landfill operations. Cover soil requires equipment, operators, maintenance and consumes airspace. Alternative cover material can only be used in certain areas and not on slopes. If 2 feet of cover soil were placed everyday that would equal 60 feet of soil every month. One quarter to one third of the airspace in a landfill is usually consumed by cover and traffic deck soil. On a permitted 30 million-ton landfill this means a $300,000,000 loss in revenue from tipping fees.

On average, it costs about $17.00 per ton to place refuse into a landfill cell in California by conventional methods. Increasing fuel costs is accelerating the rise in costs eating into profits. By pre-treating the refuse and keeping the equipment in one place, will lower costs and increase profits and at the same time be more environmentally friendly. The goal should be to lower the cost per ton to $5.00 or less. This process should extend the life of the landfill about 5 times longer than conventional methods.

The latest trend in the solid waste industry is waste-by-rail to move the waste as far as possible from cities. By using the Steam Injected Compaction Station it is the only way to load 100 tons of waste on to a standard rail car. Also by placing the compacted block into a cardboard box it is a one way trip with no deadhead to bring back an empty dirty container. This is cheaper, safer, and cleaner and saves airspace.

Waste Block Distribution

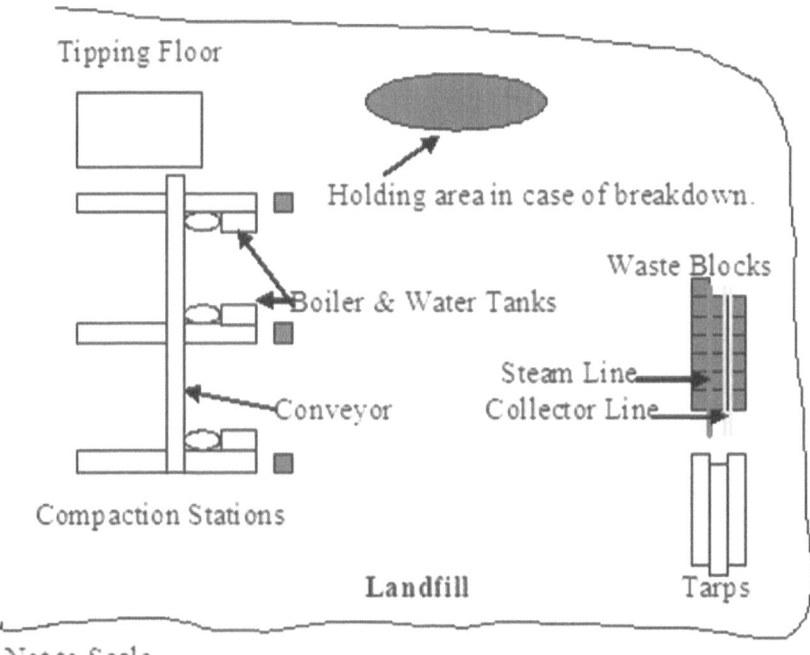

Figure 13-2

13.1 Rotary Compaction Station

Volume reduction can be used for all types of recyclables that need to be transported to a processing center. Steamed cardboard can be compacted to half the volume of dry cardboard with no apparent damage to the fibers. Twenty-five tons of steamed cardboard can be packed into a 20 foot shipping container instead of a 40 foot container.

Auto Shredder Residue (ASR) is a waste from shredding automobiles such as seats, headliner and carpeting, etc. This waste usually contains about 44% moisture, which calculates to about 5 tons of water on each truckload. Steam Compaction can remove this water and allow 5 more tons on each truck. To accommodate the production rate of the ASR coming out of the shredder a Rotary Compaction Station has been designed. This 2 cubic yard device will de-water and compact about 28 tons per hour.

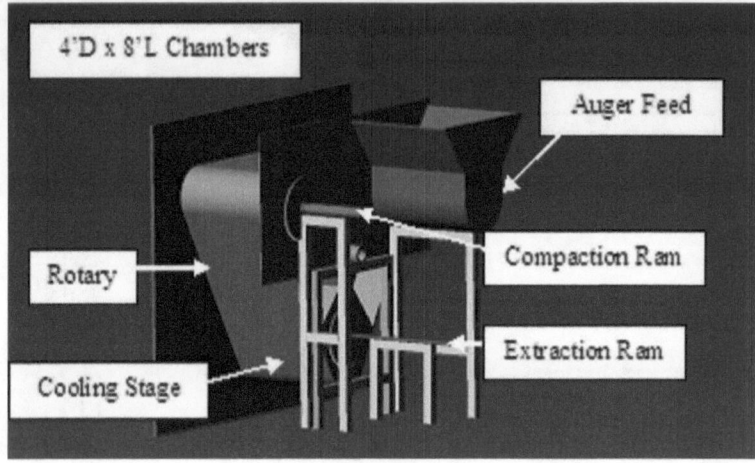

Front View
Figure 13-3

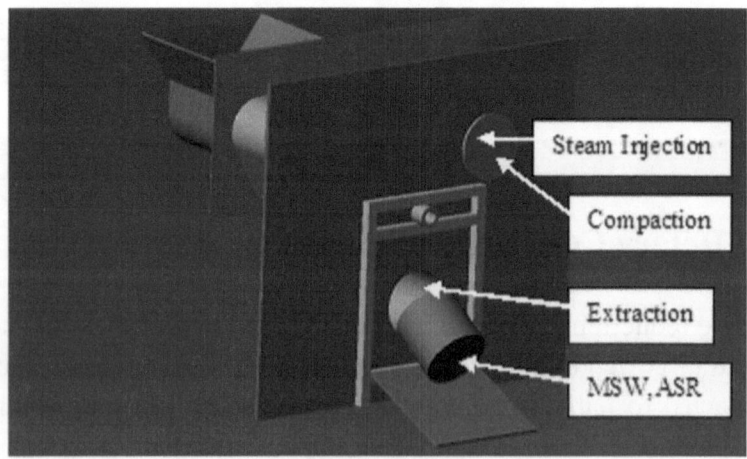

Back View
Figure 13-4

Chapter 14

Just Some Other Thoughts

In the mid 1990's I was working on various liner testing devices in the laboratory of the engineering firm I was working for. The tests involved placing various types of liner materials on a tilt table with a soil tray I constructed, that would hold various types of soil used for foundation layers etc. I then placed a 1 cubic yard box on top of the liner. The box would then be filled with various weights of lead to simulate the weight of a soil layer on top the textured liner or geotextile/liner combination, etc. The box and table were connected to linear motion transducers and an x-y plotter. The table was then tilted using an electric wench. The movements were plotted and the friction angle was determined. The soil types and moisture content could be tweaked to find the best parameters. It was a fun test.

Another test was the puncture flow test. This test bed held various types or thicknesses of liner materials with different types of punctures or cuts. The test bed made of clear acrylic could be tilted to various angles. A special water nozzle was developed to provide sheet flow over the liner. Four channels were constructed under the liner bed allowing for 4 different types of perforations in the liner. These channels would catch any leakage from the perforations and transport the water to individual reservoirs and recorded. This was a popular test with the women in the company; they would sit next to this device during their lunches and enjoy the waterfall sounds and maybe my good company as well.

One day the owner of this engineering company came into the lab and said he had a brilliant idea. "I think we can take a Claymax composite liner and use it as a top closure liner." The Claymax composite liner has a ¼" layer of Bentonite clay on one side of a liner and is usually used on the bottom of a landfill cell. The purpose is to provide another layer of protection should the liner get punctured.

The idea is that water would seep through the puncture and hydrate the Bentonite clay, which would swell and close the hole.

Something just told me that this didn't seem right. I suggested to the boss to give me some time to think about this and to do some testing before he tried selling the idea.

I simulated 2 landfill cover cross-sections in 2 aquariums with various Claymax products on top. I inserted steam nozzles into the bottom of the aquariums and connected them to a small boiler. I also installed thermocouples at various locations in the test cells to monitor the steam migration. The liner was perforated with various punctures and cuts.

The boss thought that the wet landfill gas would keep the Bentonite clay hydrated enough to prevent LFG from leaking though the liner if damaged. So to prove this theory, I injected 100 ppm methane at the bottom of the test cells until I was able to detect the methane at the perforations using a FID analyzer. I then started the steam injection into the test cell to simulate wet LFG. After a short period of time it began to work, the steam hydrated the clay and it swelled and the methane was non-detectable.

Everyone said hurray for the old man. I said, "not so fast, what happens when clay becomes saturated?" So I let the test run and an hour later the clay became saturated and the gas bubbled right though the holes. The boss said I put too much steam into the test cell. I said, "Ok we will let the clay dry out and when the methane stops leaking trough the holes I will pour a small amount of water over the holes in the liner to simulate rainwater. What do you think will happen then?" The boss looked at me for a moment, didn't say a word and walked out of the lab. I yelled after him and said sorry but that's what you pay me for.

Everyone said I had the best job in the company playing with all my toys and I must agree with them. Believe it or not at that time I still had not thought about injecting steam into landfills yet. That came after I left this company and started my own.

In 2000 I read an article in GFR Magazine about installing liners at various elevations inside landfills to improve the distribution of water in a Bioreactor.

The article went on to describe the dangers of non-uniform loading of bottom liners by the addition of liquids and listed landfills that had failed liners in Bioreactors.

I called the editor of the magazine and ask if I could submit a rebuttal to this article with an alternative solution, he said sure. To submit an article the process requires that three professionals in the industry review it. Of course the editor sent my article to the author of article I was rebutting. Well you can imagine the review I received from this person, which criticized everything about my writing skills and me but had nothing to say about the steam technology. The editor was embarrassed and apologized. Fortunately the other two reviewers gave approval so the article was published in January 2001.

After this article was published I submitted the following paper as a think piece and to invite GFR readers to respond with the best possible applications for this device. However, the editor said that the piece had to go through the prier review as well. I felt that I did not have enough empirical data to support my ideas presented in the paper. I just wanted to kick the idea around a bit before taking it to a liner company to see if they were interested. I thought that if I could show them a few letters from other professionals they would take it more seriously.

So let's try it again and see if we can have a little fun with this. Please send your comments to www.landfillengineering.com.

14.0 Magnetic Liner Anchor System

There are many types of liners and many uses for liner materials but the one thing they all have in common is that they have to be anchored in place. For example, most landfills or waste piles constructed per current regulation use an impermeable membrane (e.g. PVC or HDPE) as a bottom liner or in the cover to prevent infiltration of rainfall and thereby reduce leachate generation.

The liner is usually anchored around the edges or at key points to prevent it from moving excessively during construction or as a result of settlement of underlying waste, or during a major earthquake.

Currently, most liners are anchored by using anchor trenches, which are usually effective in most uses. However, there have been situations where liners have been disqualified for a particular project because an anchor trench could not be used because there was not enough room, soil or the slope was too steep or point load anchors would overstress the liner. Also, one of the concerns with anchor trenches and point load anchors is that they anchor the liner in place and allow for no movement of the liner itself. The anchor holds the liner at the stress points and as settlement occurs the liner begins to stretch. The system performance then relies on the designer's ability to predict the total movement and the liner materials capability to stretch without tearing or "pulling out" of the anchor. If stretch does occur, the bottom liner or the cover liner ends up with a much thinner membrane; therefore a heavier membrane must be designated in the design.

Buried anchor trenches do not reveal the loading on any particular point along the trench until signs of failure/movement are apparent and then it may be too late. The holding capacity of an anchor trench can vary over time as the soil is saturated by rain or as it dries out.

"Earth anchors" have been used in some slope applications but this requires perforating the liner with a threaded shaft to sandwich the liner between two steel plates. If it is necessary to readjust the anchor before failure occurs, the old perforation must be sealed and a new perforation made in a new location. These perforations provide a failure point in the liner and a possible location for leakage. It can be labor intensive and costly to readjust an anchor trench or an earth anchor perforation.

However, some applications may require an anchor to be forgiving as the underlying layer on which the liner is placed moves for whatever reason or there is uneven loading on the liner.

Usually, if the foundation layer under a liner moves or the loading changes and the liner moves beyond its yield, the liner must be re-anchored or the liner fails. If site conditions indicate that readjustments are going to be necessary as the foundation settles or the liner loading changes, such as in a landfill, a better alternative would be not to penetrate the liner, or bury the liner in an anchor trench until all anticipated movement has stopped.

An alternative would be to use a magnetic liner anchor. A duckbill earth anchor would still be used to anchor a steel plate to the foundation layer. Once this plate is in place the liner would be installed as usual. The liner would then be sandwiched between the anchored steel plate and a large permanent magnet. (Figure 14-1) Large permanent magnets can be shaped to provide hundreds even thousands of pounds of frictional resistance. However, this method is very forgiving and will allow sliding as the landfill or underlying foundation settles without damaging the liner.

The amount of magnetic force required is determined by the amount of area and weight of liner each anchor is required to hold in place and the angle of the slope. The magnetic resistance should not exceed the tensile strength of the liner materials used.

The thickness of the liner material and geofabrics being used will create a "magnetic air space" between the steel bottom plate and the magnet, which will be important when calculating the required pulling power of the magnets to hold the liner. The shape of the bottom steel plate is very important; it can be shaped to provide varying degrees of resistance as the magnet moves downward, such as a quadrangle. (Figure 14-2) The top of the steel plate is narrow providing less magnetic resistance on the liner allowing the liner to slip as settlement or loading occurs. As the liner and the magnet slides down it will encounter the wider portion of the steel bottom plate and increase the magnetic resistance.

This resistance should not exceed the tensile strength of the liner but will hold under higher loads/stress but will still allow the magnet to slip. The length of the steel plate can be elongated to extend the time between resetting the magnet to the top position.

Once enough movement has been detected and the magnet is half off the bottom steel plate it is a simple matter of lifting the magnet up and repositioning it over the bottom plate.

The size of the magnets are dependent on the amount of resistance required, they can be as big as 1'x 1' square or larger. A perimeter flange can be attached to the magnet to increase the force footprint on the liner. (Figure 14-3)

A bar can be attached to connect the magnets to each other and a pressure plate can be placed between the magnet points to extend their influence over a greater area. (Figure 14-4) Over time if the magnets do not perform as desired such as holding too firmly or not firm enough the magnets can be quickly changed to a different strength or size.

A properly designed permanent magnet made of the correct material can retain its magnetization for decades if it is protected from the effects of the environment (e.g. coatings) and does not experience a reverse applied field of magnitude (demagnetization).

14.1 In Conjunction with or in place of Anchor Trenches

Although anchor trenches have an established history, they may not always be applicable. The Magnetic Anchoring System (MAS) can also be used for bottom liners for landfills in place of anchor trenches and soil berms.(Figure 14-5) This is useful where space is tight (i.e. no room at edge of landfill for anchor trench) or soil is in short supply. It can be difficult to spot liner movement when it is covered with soil, which could result in liner failures. Moisture content of the soil changes the weight of the soil berm and its holding capacity.

There are some situations that require the liner to settle significantly during loading, such as a de-watering condition, which may require substantial slack in the liner prior to final anchoring. (Figure 14-5)

The MAS can allow the slack to be located above the anchoring point and the magnets are continuously reset as the liner moves into its final position, then stronger magnets can be put into place. This removes any guesswork in the final resting-place of the liner.

After several decades and no liner movement have been detected, the magnets could be removed and a top plate bolted in their place. The magnets can be reused if they are in good condition.

14.2 Construction Aids

The MAS can also be used during placement of liner and geofabric materials in windy conditions or heat expansion from the sun to hold down the liner/fabric prior to loading it. Once loading begins the magnets are removed and the bottom steel plates can be used as settlement monitoring plates.

14.3 Seismic Evaluation Tool

There have been some concerns in the past as to the effects of earthquakes on liner systems (Northridge Earthquake Report, CAIWMB). Currently, if a liner does not fail during a seismic event then the design was adequate. But if the stress was close to the failure point of the anchoring system an anchor trench may not show any indication until total failure has occurred. The MAS could be used in research studies to measure the movement of the liner during an earthquake.

The main advantage would be that the liner would not spring back and mask the amount of movement during a seismic event. If the magnet is set close to the maximum holding capacity of the liner's yield, then any deviation of the magnet can be measured after the seismic event and show the amount of movement of the liner. Laboratory tests such as a slope board on a shaker table can be developed using the MAS technique to simulate various conditions.

159

An electromagnet can be used instead of a permanent magnet so that the holding capacity can be adjusted to meet various conditions and membrane thickness. Another electromagnet can be used at the bottom of the liner to apply the simulated load on the liner.

The Magnetic Anchor System was evaluated as one of the final cover liner system options for a superfund landfill in southern California. Soil was chosen instead of geomembranes as the final cover due to slope conditions so this system was not implemented. Some laboratory studies have been performed on this method but no field studies have been attempted. It is believed that with liner/fabric materials being used in new applications everyday (i.e. landfill bioreactors, ponds, etc.) and their inherent unknowns this method may prevent some future liner failures and that additional studies on the uses of the MAS should be conducted.

The MAS may not be as cost effective as anchor trenches on very large liner systems and is not intended to replace anchor trenches, but there may be more projects available with this alternative means of anchoring liners.

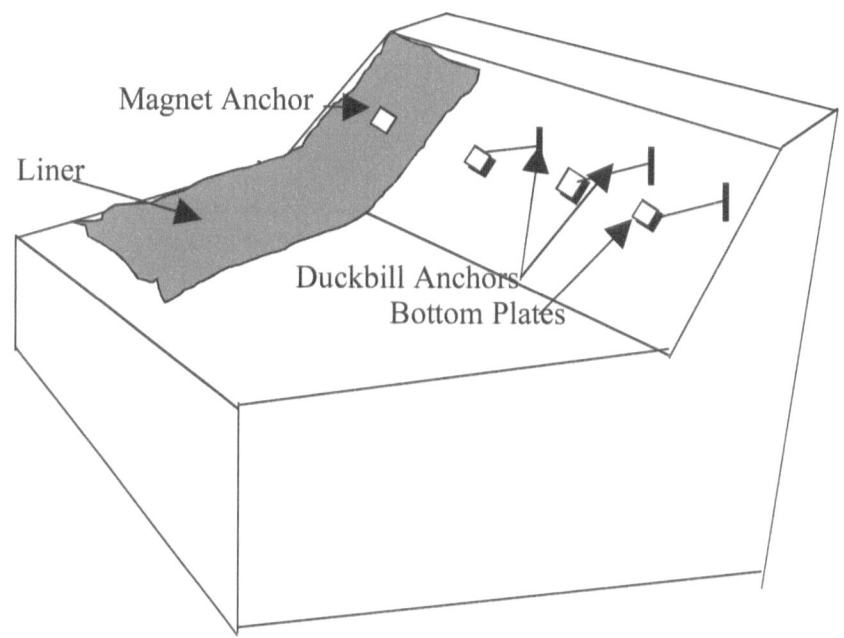

Figure 14-1

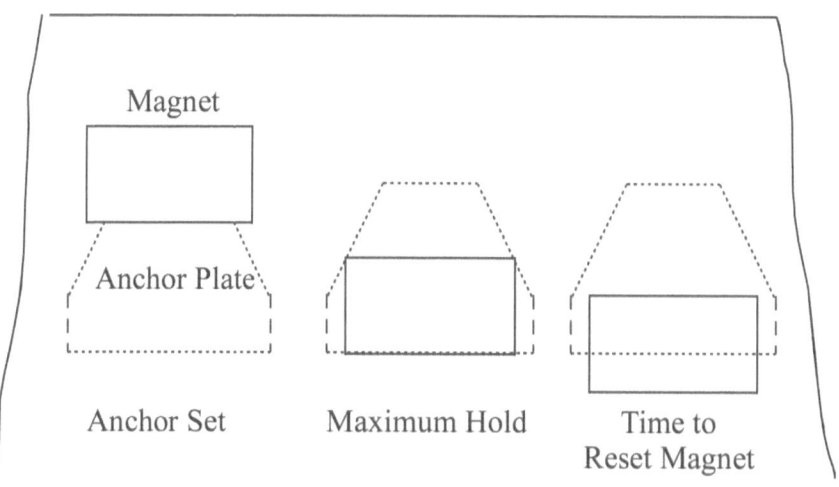

Figure 14-2

161

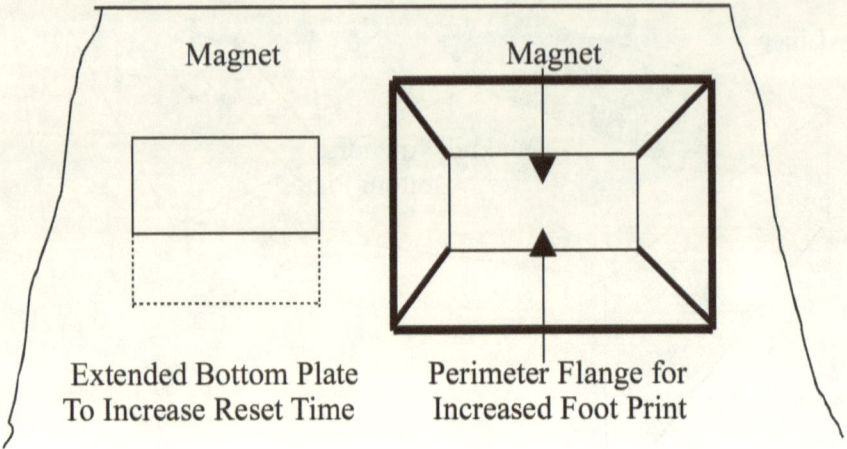

Magnet   Magnet

Extended Bottom Plate   Perimeter Flange for
To Increase Reset Time   Increased Foot Print

Figure 14-3

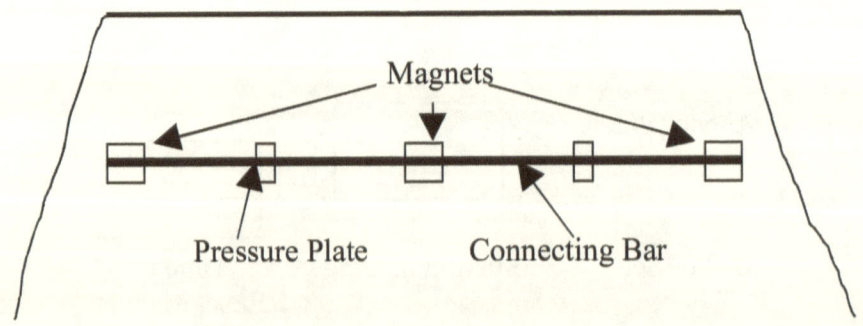

Magnets

Pressure Plate   Connecting Bar

Figure 14-4

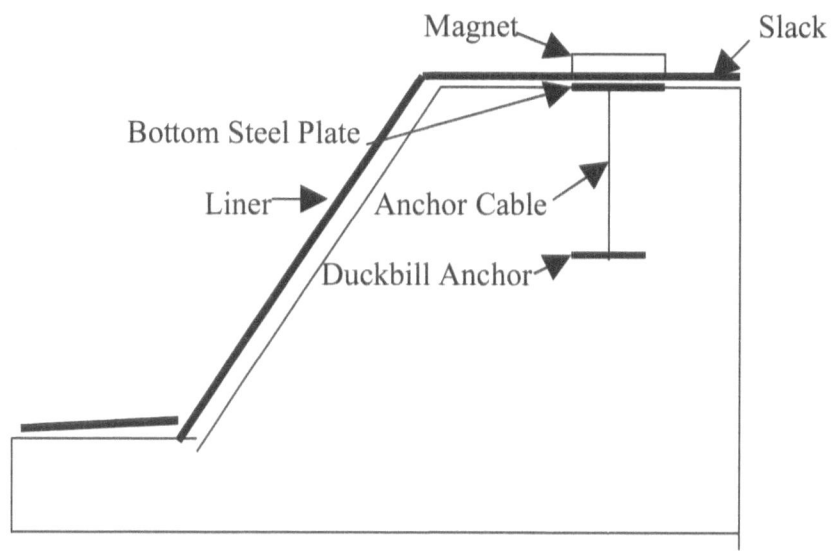

Figure 14-5

Action Items

The following list contains items that I think require action by the industry and its agencies.

- Stop drilling blindly through landfills.
- Use the PPT to locate LFG and liquid layers.
- Stop installing large diameter gas collectors.
- Check valves should be installed on all perimeter gas collectors.
- Perform Integrity Tests on all suspect Multi-Chamber Probes
- Stop installing Multi-Chamber Probes.
- The line of compliance should not be at the property line. Install Perimeter Probes within the distance from the bottom of refuse to the groundwater.
- Determine the native bottom soil type and density on unlined landfills before designing final covers.
- Install vents or gas collection on all closed landfills, give the gas a place to go instead of to the groundwater.
- Install check valves on all perimeter gas collectors.
- Stop injecting water in landfills, inject steam instead.
- Steam Injection can decompose the organic material in closed landfills in a few years instead of several decades.
- Stop using Bentonite to surface seal PPT holes, boreholes and all collectors. It is not the approved material for the final cover, the cover soil is. Bentonite in dry climates desiccates and cracks and allows leaks. The cover soil is designed not to.
- Start producing CNG, LNG, hydrogen, ethanol, methanol, e-diesel and ammonia nitrate from landfill gas.

What Now?

If you have made it this far, you have read through about 30 years of experience in the landfill and other related industries. The intent of this book was to identify shortcomings in the landfill industry and to present some possible new alternatives. How they will be received in the industry, only time will tell. You have taken the time to read this material and I am grateful. You have invested time and money to obtain this knowledge so what will you do with it now? I hope this will provide the next generation a reference for future accomplishments for improving the landfill industry. The phrase I have heard the most in this industry is, "Where has it been done before?" No one wants to be first.

One thing I have learned over the years is that there is very little courage in the landfill industry but there are many attorneys with an aversion to risk.

Another phrase I heard a lot is, "You will never get it past the agencies." So consultants and landfill owners won't even try something new. However, every time I presented a new idea to the agency with the proper supporting data, I was well received. Most agency people are the first to say that there are serious problems with the industry but they are governed by the regulations and they can't recommend to a landfill owner that something else should be done. They rely on consultants to propose alternatives but they are told "It will never get past the agencies." What's wrong with this picture?

You have to wonder if any consultants out there really want to keep a landfill in compliance, no violations no work. A simple check valve can prevent about 60% of the violations at landfills but no one is using them.

So now you have some new tools, it is now up to you to use them. Good luck.

References:

American Society for Testing and Materials, 1988, *"Method for Deep, Quasi-Static, Cone and Friction-Cone Penetration Tests for Soils"*, ASTM Standard D 3441-86.

Robertson, P.K. and Campanella, R.G., 1983, *"Interpretation of Cone Penetration Tests-PART I (SAND) and PART II (CLAY)"*, Canadian Geotechnical Journal, Vol. 20, No. 4.

Environmental Protection Agency
CFR Title 40 Part 257, 258.23

California Integrated Waste Management Board
www.CAIWMB.GOV

Robert M. Koerner, 1998
Designing with Geosynthetics, Prentice-Hall Inc.

Neal Bolten, Landfill Challenges, MSW Magazine – September/October 2007

Dr. Peter Campbell, 1998.
Basic Permanent Magnetism

Magnet Sales 2001 - Calculating the pulling force of permanent magnets

Northridge Earthquake Report, 1994
California Integrated Waste Management Board, Charlene Herbs

STI Engineering, www.landfillengineering.com

www.ingramcontent.com/pod-product-compliance
Lightning Source LLC
Chambersburg PA
CBHW032014170526
45157CB00002B/687